Commemorative Spaces of the First World War

T0188195

This is the first book to bring together an interdisciplinary, theoretically engaged and global perspective on the First World War through the lens of historical and cultural geography. Reflecting the centennial interest in the conflict, the collection explores the relationships between warfare and space, and pays particular attention to how commemoration is connected to spatial elements of national identity and processes of heritage and belonging. Venturing beyond military history and memory studies, contributors explore conceptual contributions of geography to analyse the First World War, as well as reflecting upon the imperative for an academic discussion on the war's centenary.

This book explores the war's impact in more unexpected theatres, blurring the boundary between home and fighting fronts, investigating the experiences of the war amongst civilians and often overlooked combatants. It also critically examines the politics of hindsight in the post-war period and offers a historical-geographical account of how the First World War has been memorialised within 'official' spaces, in addition to those overlooked and often undervalued 'alternative spaces' of commemoration.

This innovative and timely text will be key reading for students and scholars of the First World War and more broadly in historical and cultural geography, social and cultural history, European history, Heritage Studies, military history and memory studies.

James Wallis is a research fellow at Exeter University, and the University of Brighton. Currently employed on 'Reflections on the Centenary of the First World War: Learning and Legacies for the Future', he has worked on several post-doctoral First World War-related projects – including affiliations with the 'Everyday Lives in War' Public Engagement Centre (University of Hertfordshire) and 'Living Legacies 1914–18' (Queens University Belfast). Formerly an Arts and Humanities Research Council-funded Collaborative Doctoral Award student at Exeter and Imperial War Museums, his research explores the critical geographies of conflict heritage in a variety of contexts. Recent and ongoing projects examine the relationship between photography and conflict commemoration, and museological interpretations of the First World War.

David C. Harvey is an associate professor in critical heritage studies at Aarhus Universitiet, Denmark, and an honorary professor of historical and cultural geography at the University of Exeter (United Kingdom). His work has focussed on the geographies of heritage, and he has contributed to some key heritage debates, including *processual* understandings of heritage, extending the temporal depth of heritage, the outlining of heritage-landscape and heritage-climate change relations and the opening up of hidden memories through oral history. Along with Jim Perry, he recently edited *The Future of Heritage as Climates Change: Loss, Adaptation and Creativity* (2015).

Routledge Research in Historical Geography
Series Edited by Simon Naylor (*School of Geographical and Earth Sciences, University of Glasgow, UK*) and Laura Cameron (*Department of Geography, Queen's University, Canada*)

This series offers a forum for original and innovative research, exploring a wide range of topics encompassed by the sub-discipline of historical geography and cognate fields in the humanities and social sciences. Titles within the series adopt a global geographical scope and historical studies of geographical issues that are grounded in detailed inquiries of primary source materials. The series also supports historiographical and theoretical overviews, and edited collections of essays on historical-geographical themes. This series is aimed at upper-level undergraduates, research students and academics.

www.routledge.com/Routledge-Research-in-Historical-Geography/book-series/RRHGS

Published

Historical Geographies of Prisons
Unlocking the Usable Carceral Past
Edited by Karen Morin and Dominique Moran

Historical Geographies of Anarchism
Early Critical Geographers and Present-Day Scientific Challenges
Edited by Federico Ferretti, Gerónimo Barrera de la Torre, Anthony Ince and Francisco Toro

Cultural Histories, Memories and Extreme Weather
A Historical Geography Perspective
Edited by Georgina H. Endfield and Lucy Veale

Commemorative Spaces of the First World War
Historical Geographies at the Centenary
Edited by James Wallis and David C. Harvey

Commemorative Spaces of the First World War

Historical Geographies at the Centenary

Edited by James Wallis and David C. Harvey

LONDON AND NEW YORK

First published 2018 by Routledge

2 Park Square, Milton Park, Abingdon, Oxfordshire OX14 4RN
52 Vanderbilt Avenue, New York, NY 10017

Routledge is an imprint of the Taylor & Francis Group, an informa business

First issued in paperback 2019

British Library Cataloguing-in-Publication Data
A catalogue record for this book is available from the British Library

Library of Congress Cataloging-in-Publication Data
A catalog record for this book has been requested

ISBN: 978-1-138-12118-8 (hbk)
ISBN: 978-0-367-24524-5 (pbk)

Typeset in Times New Roman
by Apex CoVantage, LLC

Contents

PART 2

Commemorative spaces 99

Figures

Tables

Contributors

Eyal Berelovich is a PhD candidate in the Department of Middle Eastern Studies at the Hebrew University of Jerusalem. His interests lie in the military history of the Ottoman Army from the end of the nineteenth century until the demise of the Ottoman Empire, with a focus on its performance in World War I, in particular in Palestine and the Sinai Peninsula.

Paul Cornish works at Imperial War Museum London and was senior curator in the team which created the museum's First World War Galleries, which opened in 2014. He is the author of *Machine Guns and the Great War* (2009), *The First World War Galleries* (2014) and a number of papers concerning the material culture of conflict. He has co-organized five IWM-based international and multidisciplinary conferences on modern conflict and is co-editor of the resulting publications: *Contested Objects* (2009), *Bodies in Conflict* (2014) and *Modern Conflict and the Senses* (2017). Trained as an architect, landscape architect and cultural geographer.

Jeremy Foster is an associate professor in architecture at Cornell University. Both his teaching and research are informed by the cultural, bio-political and more-than-representational aspects of landscape. His articles have appeared in *Architectural Research Quarterly, Journal of Landscape Architecture, Journal of the Society of Architectural Historians, Gender, Place and Culture, Safundi* and *Cultural Geographies*, as well as edited collections *Women, Modernity & Landscape Architecture* (2015) and *Cultural Landscape Heritage in Sub-Saharan Africa* (2016).

Stefano Furlani is a graduate of the University of Trieste (Italy). He did his PhD in geomatics and GIS at the Department of Geological, Environmental and Marine Sciences of the University of Trieste (Italy). He is currently a senior researcher at the Department of Mathematics and Geosciences of the University of Trieste. His main interests include coastal geomorphology, historical geography, physical geography, geomorphology of coasts and deserts and sea-level change studies, mainly in Mediterranean areas. He projects and develops field instruments to evaluate rock surface lowering and rock erosion rates. He is the responsible for the Geoswim project *23.000 of swim surveys around the*

Mediterranean Sea. The results of his research have been published in more than 70 scientific papers since 1999 in national and international journals. He is also a science journalist.

Ritienne Gauci is a physical geographer with research interests in traditional cartography, coastal geography and geomorphology and geoheritage. She forms part of the lecturing staff of the Department of Geography at the University of Malta and has been publishing steadily in various international journals. She is national representative for Malta in the International Association of Geomorphologists and is involved in various research collaborations with other universities in Italy and United Kingdom. She is a board member on the Executive Committee of the Malta Map Society and is consulting editor of the *Malta Map Society Journal.*

Paul Gough RWA is Pro Vice-Chancellor and Vice-President of RMIT University, based in Melbourne, Australia. A painter, broadcaster and writer, he has exhibited internationally and is represented in the permanent collection of the Imperial War Museum, London, the Canadian War Museum, Ottawa and the National War Memorial, New Zealand. In addition to roles in national and international higher education, his research into the imagery of war and peace has been presented to audiences throughout the world. In addition to an exhibiting record he has published a monograph on Stanley Spencer: *Journey to Burghclere,* in 2006; *A Terrible Beauty: British Artists in the First World War* in 2010 and *Your Loving Friend,* the edited correspondence between Stanley Spencer and Desmond Chute, in 2011. Books on the street artist *Banksy* was published in 2012 and on the painters John and Paul Nash in 2014.

Paul Griffin is a lecturer in human geography at Northumbria University. He completed his PhD on 'The Spatial Politics of Red Clydeside: Historical Labour Geographies and Radical Connections' in 2015 at the University of Glasgow. His research interests include labour geography, labour history and usable pasts.

David C. Harvey is an associate professor in critical heritage studies at Aarhus Universitiet, Denmark, and an honorary professor of historical and cultural geography at the University of Exeter (United Kingdom). His work has focussed on the geographies of heritage, and he has contributed to some key heritage debates, including *processual* understandings of heritage, extending the temporal depth of heritage, the outlining of heritage-landscape and heritage-climate change relations and the opening up of hidden memories through oral history. Along with Jim Perry, he recently edited *The Future of Heritage as Climates Change: Loss, Adaptation and Creativity* (2015).

Ruth Kark, Professor Emerita at the Hebrew University of Jerusalem, has written and edited 25 books and 200 articles on the history and historical geography of Palestine/ Israel (see: http://geography.huji.ac.il/.upload/Kark/prof%20Kark. htm). Her research interests include among other topics Ottoman building, land and settlement activity; policy and law in Palestine and the introduction of modern technology and transportation to Palestine by Western powers/agents and the Ottomans, as well as Bedouin in the Middle East and Palestine/Israel; and dissent and conflict surrounding landownership in Israel (churches vs their lay communities, Bedouins vs the state, among others).

Keith D. Lilley is Professor of Historical Geography at Queen's University Belfast, and he specialises in using digital and analogue maps and mappings to explore past geographical worlds. He is currently director of "Living Legacies 1914–18", an Arts and Humanities Research Council (AHRC)–funded WW1 public engagement centre as well as the leader of a British Academy funded project called "Surveying Empires: Archaeologies of Colonial Cartography". His research is multiperiod, with particular interests in the nineteenth and twentieth centuries, as well as the European Middle Ages (c.900–1500). His books include *Urban Life in the Middle Ages* (2002), *City and Cosmos – the Medieval World in Urban Form* (2009), *Mapping Medieval Geographies* (2013), and presently he is writing a new book called *Realms of Rule: Survey, Statecraft and Sovereignty under the Plantagenet Kings*.

Raphael Mizzi works at the University of Malta and coordinates the Green Travel Plan within the Institute for Climate Change and Sustainable Development. He first graduated in 2010 in BSc business and computing, followed by GIS (geographic information systems) in 2012. He has a keen interest in WW1 and WW2 research and memorabilia, and is an active member in the Malta Archaeological Society.

Brian S. Osborne is Professor Emeritus of Geography at Queen's University, Kingston, Ontario, Canada, where he has taught since 1967. His research areas include aboriginal history, settlement history, cultural landscapes and the role of the "culture of communications" in the development of a Canadian sense of place. He has published extensively on the Kingston area, while his current research considers symbolic landscapes, monumentalism and performed commemoration as contributors to the construction of social cohesion and national identity. His current projects are *Constructing Canadian Identity in a Trans-national World*, *Canadian Utopian societies* and *Co-memorizing the "Great" War*.

Catriona Pennell is a senior lecturer in history at the University of Exeter. She specialises in the history of nineteenth and twentieth century Britain and Ireland with a particular focus on the social and cultural history of the First World War and British imperialism in the Middle East. Her most recent research has explored the relationship between war, education and memory and has led to an employment of interdisciplinary methodologies, particularly around the boundaries of history, politics and geography. Her AHRC-funded 'The First World War in the Classroom: Teaching and the Construction of Cultural Memory' has led to her appointment as academic lead on the government-funded

First World War Centenary Battlefield Tours Programme, leading their evaluation of pupil experience between 2015 and 2019.

John A. Schembri is a professor in geography at the University of Malta and lectures in human geography with particular reference to the Mediterranean. As a coastal geographer, his interests are in development in ports and harbours and historical heritage along urban coastal areas. The latter includes research on early twentieth century ordnance survey cartography of the Maltese Islands. He is a regular contributor to courses organised by the International Ocean Institute in Malta.

Shanti Sumartojo is a vice-chancellor's research fellow in the School of Media and Communication at RMIT University, Australia. She has published on public memorials and the 'atmospheres' of commemorative sites, sensory perception of design and the built environment, how people encounter and make sense of digital data in their everyday lives and the impact of creative practice on public space, including collaborating on performance and built installations. She is the author of *Trafalgar Square and the Narration of Britishness, 1900– 2012* (2013); and co-editor of *Nation, Memory and Great War Commemoration: Mobilizing the Past in Europe, Australia and New Zealand* (2014) and *Commemorating Race and Empire in the Great War Centenary* (2017).

James Taylor is Assistant Director of Narrative and Content at Imperial War Museum London. He led the curatorial team who worked on the First World War Galleries that opened on site in 2014 and has authored a chapter within Nicolas Saunders and Paul Cornish's edited collection *Bodies in Conflict* (Routledge, 2014).

James Wallis is a research fellow at Exeter University, and the University of Brighton. Currently employed on 'Reflections on the Centenary of the First World War : Learning and Legacies for the Future', he has worked on several post-doctoral First World War–related projects – including affiliations with the 'Everyday Lives in War' Public Engagement Centre (University of Hertfordshire) and 'Living Legacies 1914–18' (Queens University Belfast). Formerly an Arts and Humanities Research Council-funded Collaborative Doctoral Award student at Exeter and Imperial War Museums, his research explores the critical geographies of conflict heritage in a variety of contexts. Recent and ongoing projects examine the relationship between photography and conflict commemoration, and museological interpretations of the First World War.

Ross Wilson is a senior lecturer in modern history and public heritage at the University of Chichester. He has written on the experience, representation and memory of the First World War in Britain and the United States. His wider research focuses on issues of museum, media and heritage representations in the modern era. This work has been published in the books, *Representing Enslavement and Abolition in Museums* (2011), *Landscapes of the Western Front* (2012), *Cultural Heritage of the Great War in Britain* (2013), *New York* *and the First World War* (2014), *The Language of the Past* (2016) and *Natural History: Heritage, Place and Politics* (2017).

Xu Guoqi received his PhD in history from Harvard University and currently is a full professor of history at the University of Hong Kong. He is author of the following recent books in English: *Asia and the Great War: A Shared History* (2016); *Chinese and Americans: A Shared History* (2014); *Strangers on the Western Front: Chinese Workers in the Great War* (2011); *Olympic Dreams: China and Sports, 1895-2008* (2008); *China and the Great War: China's Pursuit of a New National Identity and Internationalization* (2005). Professor Xu is currently working on a book manuscript tentatively titled "Idea of China" which is under contract with Harvard University Press.

Foreword by the series editors

We are delighted to have *Commemorative Spaces of the First World War* published in the Routledge Research in Historical Geography series. The volume is exemplary for a number of reasons. First, it brings together an exciting mix of new and established scholars from the United Kingdom, the USA, Denmark, Israel, Malta, Canada and Australia, to consider the commemoration of the First World War from a range of international perspectives. Second, the volume provides an innovative critical geographical template for studying and understanding the First World War, which really sets this volume apart from others on the conflict. Individual chapters consider contested spaces of memory, landscapes of protest, cartographic representations, sites of remembrance and commemoration, warscapes and battlefield spaces. This is the first such book-length collection of works by historical geographers on the First World War and it will make a significant contribution to wider debates within human geography, heritage and museum studies and cultural and military history. Third, the volume is very timely. The publication of *Commemorative Spaces of the First World War* will coincide with the First World War Centenary and, in particular, with the Centenary of the Third Battle of Ypres, which took place over the summer and autumn of 1917 and produced many of the most famous photographs of trench warfare. Finally, the volume makes an excellent addition to the growing catalogue of books in the Routledge Research in Historical Geography series. We feel sure that it will play a significant role in introducing historians and others to the value of a historical-geographical approach to the study of war, heritage and memory.

Simon Naylor and Laura Cameron

Preface

Centennial reflections

A cataclysmic event with an immeasurably long shadow, the First World War was, and remains, an inherently spatial conflict. Accounting for the scale of its meaning over the decades that have followed represents a difficult task. The focus of this volume is to reflect upon what geography – particularly historical geography – has contributed to how the First World War can be understood within the contemporary context of its centennial commemorations.

The conflict has attracted increased and notable analysis from geographers, predominantly since the mid-1990s. Such work has done much to influence our understanding of the broad concepts of national identity, (local and individual) remembrance and memorialisation, and has thus provided new insights into the relationship between warfare and space. In a desire to gather the latest thinking on these broader notions, this edited collection originated from three sessions of papers given at the Royal Geographical Society (with IBG) Annual Conference in August 2014. This was situated in the midst of the opening commemorations for the First World War Centenary, with the sessions held in partnership between the Royal Geographical Society and Imperial War Museums (IWM). The intention was simultaneously to showcase and explore geography's continuing contribution to our collective understanding of the conflict at this time of unparalleled prominence within both the academic and public spotlight, as popular, policy and scholarly interest reached new heights. Thus the value of the work presented at these sessions – the ideas put forward from an international and interdisciplinary group of geographers, historians and heritage sector practitioners – formed a conceptual contribution that warranted collation and distribution amongst wider audiences. Furthermore, this preface, alongside the introduction that follows in Chapter 1, sets out a guiding thread that spans the volume, in arguing for a fresh take on how the war is understood today. Whilst chapter contributions herald from a range of disciplinary contexts, all draw upon a critical geography perspective to interrogate some of the practices and processes of remembrance, and what this means for understanding the conflict today and beyond.

Bringing this book to fruition has been the result of extended support and goodwill from a number of individuals and organisations. The partnership between the

Royal Geographical Society and Imperial War Museums provided the forum for stimulating discussion, both in the formal conference sessions and on the connected field trip to the newly opened First World War galleries at IWM London. We would like to thank everyone involved in the sessions, both audiences and contributors, particularly those not able to publish their research within these pages, but whom nonetheless greatly contributed to their overall success.

We owe a great deal to the chapter contributors, who (appeared to) put up with our sporadic nagging e-mails and occasional tardy responses with equal good humour. While the majority of the chapters within this book featured at the RGS sessions, we also commissioned further chapters to fill perceived omissions. Some gaps will inevitably remain, however, and it is very much our view that this present volume can only ever provide a partial coverage of the field. While we accept responsibility for such shortcomings, we also feel that such criticisms should be seen as a testament to the vibrancy of this broad and interdisciplinary field of study. Finally, we would like to thank Mary Hilson and Susie Golics, both of whom have given unstinting support and advice to the editors during the volume's compilation.

James Wallis
David C. Harvey

1 Introduction

Conflicting spaces – geographies of the First World War

James Wallis and David C. Harvey

> The remembering and forgetting of war is not an object of disinterested enquiry but a burning issue at the very core of present day conflicts over forms of the state, social relations and subjectivity.
>
> (Ashplant et al. 2000: 6)

In recent years, numerous historical and cultural geographers have focused upon the spatial elements of identity, belonging and control within the context of the First World War.[1] As well as forming a distinctive intervention within First World War studies, much of this work has also been influential within broader fields of enquiry into, for instance, iconography, landscape representation and memorialisation. Alongside this work by geographers, and similarly responding to the cultural turn within the academy, a number of First World War scholars have drawn upon geographical approaches to make critical interventions, for instance, towards our understanding of national representation and imperial identities. Thus articles by Gough (2004) and Foster (2004) both analysed First World War battlefield sites and their role in producing cultural memory within national and imperial contexts. Furthermore, there is a *geography* to the range and focus of this scholarly activity, particularly concerning the examination of collective memory and remembrance activity. While different sites of the Western Front have tended to be central in much scholarly output, the analysis of how the mythologised landscapes of Gallipoli have been mobilised has formed a central plank of the examination of Australian collective memory.[2] There has also been distinctive focus upon Ireland in articulating the past and present roles of landscape and memory within that contextual setting.[3] Drawing upon examples from the Second World War, it is evident that countries remain entangled in representing an imagined nation, both spatially and temporally, with narratives 'elevating and naturalizing certain elite memories whilst marginalizing others' (Muzaini and Yeoh 2007: 1289). Authors have revealed war memorials as dynamic inscriptions of memory onto space – setting dominant socio-spatial relations in stone, as they recall and represent selective histories that hide as much as they reveal. Indeed, within a 1914–1918 context the legacy of war memorials has provided much inspiration for an interdisciplinary group of academics.[4] Moreover, the commemorative practices that emerged out of this global event 'have endured throughout the twentieth century . . . the motif of individual sacrifice for the national cause has remained central to the act of national commemoration' (King 2010: 7). In essence, the crossover between different scales and interactions between the global through to the local, are increasingly being reassessed via the use of spatial frameworks.

Historical geographies of the First World War, therefore, can provide a way in which to rethink the spaces of the conflict, and in particular, the palimpsest-like nature of battlefield sites (principally the former Western Front, located in France and Belgium) and the relation between so-called battle-fronts and home-fronts. The potential for innovative ways to re-engage – even re-interpret – our broader understanding of these landscapes is something that historical geography has the epistemological tool kit to undertake. Similarly, a major strength of the sub-discipline's practitioners lies in their desire to undertake critical scholarship that questions the past in relation to its meaning in the present, and to dissect the interlocking relationships between time and space. Within the context of a post 'living-memory' landscape, now is the time to promote such a message – advocated in the knowledge that it has, thus far, largely been those scholars based within heritage studies and history, who have provided critical commentary on the contemporary resonance of the centenary.

Reflecting on the centennial moment, Ziino (2015: 1) recognised that this event 'has naturally been a source of considerable debate and stimulus – at least among academics and politicians, and in cultural institutions – for a long time before its realization in 2014 and beyond'. As commemorations progress towards an inevitable climax, new academic research from an array of disciplines continues to illuminate diverse aspects of this global conflict, alongside its ensuing legacies (Sumartojo and Wellings 2014; Ziino 2015; Drozdzewski et al. 2016). Specifically within the UK context, the government's ongoing commemorative programme continues to develop connections, and sustain new interactions, between the academy and public engagement projects.[5] The timing is, therefore, opportune to reflect upon this imperative for collaboration, to take stock of what has been garnered from these activities in time to contemplate what might lie ahead for the remaining anniversary period and particularly its legacy in the twenty-first century.[6] The confines of this volume subsequently build upon the understanding set out by Gegner and Ziino (2012: 4) that the physical sites and spaces of the First World War have acquired greater claims to authenticity in light of the passing of living memory of the conflict. Accordingly, we wish to account for the ways in which this burden of commemoration has been enacted in practice, whilst additionally assessing its reception and affective agency amongst audiences. In this way, the volume reacts to the broad, interdisciplinary and immanent conversation that is currently reconsidering the practices, processes and poetics of remembrance, situated amidst an ongoing critical appraisal of historical conflict studies.

Gegner and Ziino (2012: 1) have observed, 'If heritage can be understood as the selective use of the past as cultural and political resources in the present, then there are few fields more productive for understanding that process than the heritage of war'. To break this down further, one must gauge the core interlinking strands that formulate conflict heritage; notions of memory and the construction or maintenance of identity narratives stand inherent within such debates. Drozdzewski et al. (2016: 3) have already referred to the undoubted 'centrality of championing a nation's collective identity' amidst the centennial commemorations, whilst Mycock et al. have documented an intertwining of the personal and public, 'subject to both family and state rituals that link the individual to the nation' (2014: 2). Arnold-de Simine (2016) and Wallis (2015) have examined how the influence of digital media is changing the ways in which commemorative practices are enacted, and in the process made familiar, to an array of public users. So if, as Drozdzewski et al. attest, memory is 'part of a tripartite relationship with identity and place' (2016: xiii), then we must consider the importance of paying attention to how different scales of experience work in practice. This is especially relevant in an age of social media, but moreover amid an era of existential crises – both for European identity and for the United Kingdom and its constituent parts. Indeed, it has been interesting that the centenary period has also witnessed referenda on Scottish independence and the United Kingdom's withdrawal from the European Union, as well as ongoing crises concerning refugees, migration and growing national populist movements across Europe. Questions about how commemoration of the First World War inflects questions of belonging within the national, continental, or even global context (see, for instance, Mycock et al. 2014: 3) require a response from geographers. Indeed, thus far geographers have not commented extensively on the First World War Centenary, despite the conflict's extraordinary hold upon the British collective imagination (though see Harvey 2017).

The conflict was undoubtedly emblematic, in terms of both its industrial nature and the sheer number of casualties, beyond constituting an unprecedented experience for people and governments alike. This explains the unparalleled levels of popular and academic interest that have followed, but particularly in the build-up to, and during its centenary.[7] Academic attention has also reviewed the contemporary remembrance role of the conflict within a British context (Andrews 2011; Pennell 2012b; Jones 2014). This literature has tended to focus on debates around the politics of commemoration, and the challenges to 'establishing "national" narratives and memory cultures to mark the First World War Centenary that are inclusive and yet recognise diversity in how the conflict is remembered across Britain and across its former empire' (Mycock 2014a: 153, 161). From this, we see war commemoration enduring as

> primarily a political project whereby the state and its institutions mediate and order formal and informal collective memories and histories. The promotion of a homogenous national identity that references important conflicts is seen to establish symbolic continuity between the past, present and future of a nation-state.
>
> (Mycock 2014a: 154)

Amidst a desire to scrutinize these channels of commemoration, we tasked our authors with an agenda to account for the ways in which political decisions have informed both the production and renewed attention upon these commemorative sites and spaces. It is worth labouring the point that

> the struggles over war memory remind us that the narratives attached to the First World War are not static, or agreed, but are subject to constant contestation, and change over time. This is in the nature of cultural memory, and in recognizing this, we can see the life histories of remembering, at a series of levels – public, private, institutional – and the cultures of remembrance that those processes have bequeathed to the present.
>
> (Ziino 2015: 1)

We are consequently inclined to view the current devotion to commemoration as something more than a sense of anniversary-driven responsibility. An appropriate framework to add weight to this conviction is outlined within Sørensen and Viejo-Rose's (2015) work, which connects discussions around memorialisation to specific locales, in tracing processual shifts in their meaning over time. In particular, it demonstrates their notion that 'places are not just "the heritage of war" but actively participate in the recovery and remaking of communities' (Sørensen and Viejo-Rose 2015: 1). By utilising their geographically informed concept of a 'biography of place', their edited volume offers a number of case studies that delve into how a meaning of place, positioned within particular locales, is continually remoulded in the present and by whom. Emphasis is assigned towards illuminating these practices and mechanisms; 'Heritage sites have particular agency for they are effective means of creating narrative links between people, their pasts and their surroundings' (Sørensen and Viejo-Rose 2015: 8). Additionally, therefore, the present volume offers a response through its collective empirical findings to those ideas about memory, place and identity raised by Drozdzewski et al. (2016). In stressing the importance of paying attention to the *encounters between*, Drozdzewski et al. (2016, emphasis in the original) prompt us to explore how memories of war and conflict manifest physically, spatially and temporally in place. This volume therefore, seeks to offer insight into questions around spatial dimensions of belonging and memory – showing these as unfixed, ever-changing and seemingly contradictory entities. It both considers the issue of whose duty it is to remember and provides critical nuance to the question of what is remembered and why. In so doing, the volume perceives the terrain of First World War commemoration as an inherently active space. Testifying to the production, management and politics of these commemorative spaces, therefore, its chapters collectively sketch the interplay between agents of memory and heritage, exploring the processes and implications of transmission, as well as documenting pockets of resistance. The sum of its individual parts therefore, offers a dynamic yet novel investigation into the function and guises of commemorative spaces of the First World War (as located within a specific temporal context), thereby widening our existing understanding.

The First World War often seems to be about (static, stable and indelible) *lines*, of trenches, barbed wire, western and eastern fronts, of a well-known chronology, from an event in Sarajevo in July 1914, to a precise moment in the late morning of the 11 November 1918. This linear chronology has become ritualised through ceremony ever since, now further cemented through centennial public reflection. This volume provides the space to rethink what we know about these lines, challenging some of these linear aspects of First World War understanding and providing a platform for new thinking on well-trodden ground.

Outline of the collection

The collection is formally divided into two sections, via interconnecting themes. In Section 1, entitled Rethinking, and Looking beyond the Front Line, the volume provides room for new historical-geographical examination of the First World War itself, exploring new and critical conceptualisations of the spaces of the battlefield and specifically going beyond the standard Western Front lens of enquiry, to reveal and explore other fighting fronts and lenses of analysis at different scales. Moving from the familiar terrain of the Western Front, these chapters take an innovative approach, adding distinct layers to this field of enquiry. Thus, the section showcases work that examines the war's impact in more unexpected locations and subsequently blurs the boundary between home and fighting fronts. Focusing on the perceptions and experiences of the war both within less-travelled terrain of anti-war activity on the Home Front and within less-studied contexts (Malta, the Middle East and Asia), this section both investigates the experiences of the war among civilians and often overlooked combatants.

Focusing on geographical interests in representation and 'geo-visualisation', Chapter 2 (Gough) examines the artistic representation of memoryscapes. Gough examines the spatiality of the Western Front from the perspective of those artists and surveyors charged with lending visual form to the battlefield chaos. Examining how these individuals responded to vistas of seemingly featureless prospects – and the prospects of apparent features such a notable trees – Gough considers how the requirements of accuracy, calibration and standardisation were negotiated. The situation required a different way of knowing, ordering and navigating spaces of military violence; an interplay of time and space in the context of the constant mutability of the apparently static space of the front line; something that, as the author makes clear, has resonance with contemporary spaces of conflict in the twenty-first century.

While there has been a good deal of output on the Home Front in recent decades, Chapter 3 (Griffin) takes this strand of work in an alternative direction through exploring the anti-war political scene in the Clydeside region of Scotland during the conflict. Using the voices of three key activists, Griffin examines the relationship between the politics of class with different strands of anti-war philosophy and the relationship between the local scene on Clydeside and wider translocal activity, noting the complexity of motivations for contesting the First World War in different scalar contexts. The chapter suggests that the multiplicity of experiences of war and anti-war are mirrored by multiple forms of commemorative activity today.

The authentic voices of civilian commentators also feature prominently within Berelovich and Kark's chapter (Chapter 4), which takes the reader beyond the terrain of the Western Front to explore civilian experiences of the conflict in Palestine. Using contemporary diaries and memoirs, events as they unfolded are traced through the complex interplay of individual narratives of daily life and the crucial, but mutable, relationship with the Ottoman state and to the wider conflict. Berelovich and Kark ably demonstrate the importance of understanding how knowledge travels, with changing degrees of confidence and detail – providing crucial insight into how the war was perceived and experienced by civilians in a region that is often elided in standard accounts.

Using an array of contemporary published reports and newspapers, together with a series of maps, Schembri, Gauci, Furlani and Mizzi (Chapter 5) document the largely ignored setting of the islands of Malta. Away from the front line, but very much at the centre of activities to support the war effort in the Mediterranean and near East, the experience of Malta has certain 'Home Front' familiarity, as the civilian population became further enrolled in the victualing, reconnaissance and support services necessary for the prosecution of an increasingly global war. The strategic location of the islands, however, together with the sheer scale of its medical and hospital functions – injured and convalescing troops added more than 50 per cent to the total population by 1918 – provide a distinctive and place-specific element to the Maltese experience, which provides an instructive intervention in our understanding of this global conflict.

Finally, in this section, Xu Guoqi (Chapter 6) compels us to take note of the effects and ramifications of the conflict beyond Europe and the near East. His comparative postcolonial analysis of East and South Asian experiences of the First World War provides a welcome transnational perspective to the collection, as well as an edifying reminder to look beyond the Eurocentric aspects of the conflict. The war was seen as an opportunity for many nationalists and their various causes within Vietnam, India, Japan and China, but while experiences of those caught up in the conflict differed, a shared sense of disillusionment can be identified as a key moment in the histories of these countries.

The second substantive section of the collection, entitled Commemorative Spaces, focuses on the theme of commemoration within a contemporary context, thereby offering a historical-geographical account of how the First World War has been memorialised. This includes an examination of remembrance processes operating within 'official' spaces of commemoration, before contemplating the mood of centennial reflection and affect within different publics. Scrutinizing both official and alternative spaces of commemoration, one of the key themes of this section acts to historicise the practice of commemorative activity around the world. Furthermore, this section also critically evaluates the politics of hindsight in the post-war period; what commemoration has done and the potential of what it could yet do.

The section opens with a critical account by Wallis and Taylor (Chapter 7) of the redevelopment of the Imperial War Museum's First World War galleries in 2014. In particular, a close analysis of the political and museological contexts of presentation within the galleries is pitched into a broader understanding of how the First World War is remembered and displayed within the post-'living-memory' era. Functioning as a national war museum, the chapter illuminates understanding into the processes, agency and constraints that govern the messages transmitted at this site, via an eyewitness account that delves into the politics of its displays.

More than 33 million battlefield maps were made by British cartographers during the First World War, which saw ordinary soldiers become familiar with using and producing maps. In a contemporary reflection of this notion of the citizen cartographer, Lilley's chapter (Chapter 8) investigates the experiences and products of amateur scholars today who are involved in digital cartographic research. While the First World War witnessed a burgeoning of the general public's aptitude and appetite for maps, popular cartography projects today have seen digital mapping become enrolled in public commemorative practices across a multitude of communities.

Investigating the broad historical terrain of transnational commemoration within white settler nations, Foster (Chapter 9) explores their transition from key locations of emergent national pride to more nuanced and hybrid spaces of memory. While the memorials – on the surface – seem little changed, the nature of remembering, together with the motivation and the actual encounter have undergone a significant transformation. Foster finds that thinking with landscape provides a deeper time-consciousness and enables a greater appreciation of an affectively rich account of transnational battlefield memorial spaces.

Turning to the centenary events at the national Australian Shrine of Remembrance, Sumartojo (Chapter 10) contemplates the changing meaning of Anzac memorial experiences and activities. In providing room for the investigation of the key centenary events at the Shrine within the context of tourist consumption, educational pedagogy and political affectivity, Sumartojo complicates the long-standing standardised role of national commemoration. Her findings add welcome insight into the complexity and malleability of contemporary Australian national identity. Building upon these discussions around the pedagogic role of commemorative activities, Pennell (Chapter 11) assesses the changing meaning of battlefield visits to the Western Front, particularly through the themes of the traveller as tourist and as pilgrim. Pennell charts an evolution of battlefield visits, especially following the increasing prominence of such tours since the 1990s and specifically interrogates the meaning and experiences of officially organised tours by British schoolchildren on the occasion of the centenary.

Remaining within the national context, but duly focussing on 'new' commemorative practices and locations, Wilson (Chapter 12) considers how alternative spaces relate to existing spaces of commemoration in a complex interplay of national, communal and personal acts of remembrance. Drawing on nonrepresentational theories, and examining a wide variety of sites across Britain, Wilson notes how centennial acts of commemoration entreaty an individual to bear witness and the apparent burden of remembering, in terms of what this means for contemporary audiences. This theme is progressed in Chapter 13 by Osborne, who provides a reflective account of how this *Great* war has been remembered over the course of the last century. Reflecting on a series of *scapes*, from the realities of warscapes, the literary and artistic rendition of inscapes and the changing experience of memoryscapes, Osborne vocalises the need for a more purposeful embrace of a new ethics of conflict through which the energizing force of memory can promote interventions to attain a better future.

Finally, Cornish (Chapter 14) brings together the volume's collective ideas into a piece that examines the concept of commemorative spaces as a significant reflective overview. In particular, Cornish documents a variety of national obligations to remember, to understand how the events they depict are far from fixed, yet continually reworked within the present. Cumulatively, the chapters assist in recasting understanding on the subject of commemorative spaces beyond their more traditional and normative conceptions. In distinguishing official and unofficial forms of commemoration, the overall findings help advance empirical debates, developed through intersecting discussion. Critical commentary on the management and interpretation of specific sites showcases the agents and enactors of remembrance, as well as the processes behind the constructed nature of commemoration. In using the lens of historical geography, its value therefore lies as a conceptual contribution towards new knowledge of First World War heritage.

Notes

1 For example, see Graham (1994), Heffernan (1995), Clout (1996), Morris (1997), Johnson (1999, 2000: 255–258, 2003), Graham et al. (2000: 36–40) Gough and Morgan (2004), Hoelscher and Alderman (2004), Dwyer and Alderman (2008), Wilson (2011), Sumartojo (2016a). For a more detailed breakdown of work completed by historical geographers focusing upon the First World War, see McGeachen (2014).
2 See Hoffenberg (2001), Slade (2003), Scates (2006, 2009, 2016), Ziino (2012), Frame (2016). For broader works investigating the national Anzac concept, see Fenton (2013), Bongiorno (2014), Mckenna (2014), Wellings (2014), Sumartojo (2015, 2016b) and for the conflict's impact upon Canada specifically, consult Vance (1997). For works redefining the conflict as a multi-racial struggle, and that examine colonialism via the interplay between Dominion and Empire, see Smith (2004), Sheftall (2009), Das (2011), Killingray (2012), Bourne (2014), Olusoga (2014), Basu (2015), Maguire (2016), Walsh and Varnava (2017), Sumartojo and Wellings (Forthcoming).
3 For more information, see Jarman (1999), Johnson (1999, 2003), Graham and Shirlow (2002), Switzer and Graham (2010). Work on Irish commemoration with its relationship to Britain reveals complex issues of multiple heritages and nationalism (See McCarthy 2005, 2013; Morrissey 2005) and demonstrates 'the importance that historical events play in the politics of contemporary Ireland' (Jarman 1999: 171). For discussion within the centennial context, see McAuley (2014), Grayson and McGarry (2016).
4 The topic of First World War memorialisation has been well documented. Key works include Moriarty (1995, 1997, 1999), King (1998, 1999), Connelly (2002), Gordon and Osborne (2004), Black (2004), Inglis (2005), Stephens (2007), Macleod (2013).
5 Over the course of 2014–2019, the Arts and Humanities Research Council is funding five national Engagement Centres, designed to link up academics with community groups

interested in researching aspects of the First World War. Outcomes have offered nuanced insight principally upon the subject of Britain's Home Front, thereby reflecting a recent public appetite for social history (see Adie 2013; Charman 2014; Andrews and Lomas 2014). The BBC and the Heritage Lottery Fund similarly are similarly operating large-scale funding streams for related projects and activities during the conflict's centennial commemorations.

6 Whilst popular interest in the event seemed to wane after the enthusiasm of August 2014, the aforementioned innovative works on the academic front range from social and cultural history to politics, material culture studies and twentieth century literature. Indeed, the abundance of recently published works has included a significant strand of scholarship that examines the topic of war commemoration more broadly. For instance, see Drozdzewski et al. (2016), Reeves et al. (2016), West (2016), Muzaini and Yeoh (2016).

7 Much has been written on the subject, including works on its cultural memory (Mosse 1990; Parker 2009), remembrance within a family framework and its implications (Winter 1999; Todman 2008, 2009; Holbrook and Ziino 2015), the material culture of the War (Saunders 2004; Saunders and Cornish 2009), its archaeology, (Wilson 2007; Miles 2016) and gendered experience (Meyer 2009). Overview monographs have been produced by numerous historians (Strachan 2003; Stevenson 2005; Todman 2005; Pennell 2012a; Reynolds 2013; Wilson 2013; Winter 2014).

Bibliography

Adie, K. (2013) *Fighting on the Home Front: The Legacy of Women in World War One*, London: Hodder & Stoughton.

Andrews, M. (2011) 'Mediating remembrance: Personalization and celebrity in television's domestic remembrance', *Journal of War and Culture Studies*, 4(3): 357–370.

Andrews, M. and Lomas, J. (eds.) (2014) *The Home Front in Britain: Images, Myths and Forgotten Experiences since 1914*, London: Palgrave Macmillan.

Arnold-de-Simine, S. (2016) 'Between memory and silence, between family and nation: Remembering the First World War through digital media', in A. Dessingué and J. Winter (eds.), *Beyond Memory: Silence and the Aesthetics of Memory (Routledge Approaches to History)*, London and New York: Routledge, 143–161.

Ashplant, T., Dawson, G. and Roper, M. (2000) 'The politics of war memory and commemoration: Contexts, structures and dynamics', in T. Ashplant, G. Dawson and M. Roper (eds.), *The Politics of War Memory and Commemoration*, London and New York: Routledge, 3–85.

Basu, S. (2015) *For King and Another Country: Indian Soldiers on the Western Front, 1914–18*, New Delhi: Bloomsbury.

Black, J. (2004) 'Ordeal and re-affirmation: Masculinity and the construction of Scottish and English national identity in Great War memorial sculpture 1919–1930', in W. Kidd and B. Murdoch (eds.), *Memory and Memorials: The Commemorative Century*, Aldershot: Ashgate Publishing Limited, 75–91.

Bongiorno, F. (2014) 'Anzac and the politics of inclusion', in S. Sumartojo and B. Wellings (eds.), *Nation, Memory and Great War Commemoration: Mobilizing the Past in Europe, Australia and New Zealand*, Bern: Peter Lang Academic Publishers, 81–97.

Bourne, S. (2014) Black Poppies: Britain's Black Community and the Great War, Stroud: The History Press.

Charman, T. (2014) *The First World War on the Home Front*, London: André Deutsch in association with Imperial War Museums.

Clout, H. (1996) *After the Ruins: Restoring the Countryside of Northern France after the Great War,* Exeter: University of Exeter Press.

Connelly, M. (2002) *The Great War, Memory and Ritual: Commemoration in the City and East London, 1916–1939,* Suffolk: The Royal Historical Society & The Boydell Press.

Das, S. (ed.) (2011) *Race, Empire and First World War Writing,* Cambridge: Cambridge University Press.

Drozdzewski, D., De Nardi, S. and Waterton, E. (eds.) (2016) *Memory, Place and Identity: Commemoration and Remembrance of War and Conflict,* Oxford and New York: Routledge.

Drozdzewski, D., De Nardi, S. and Waterton, E. (2016) 'The significance of memory in the present', in D. Drozdzewski, S. De Nardi and E. Waterton (eds.), *Memory, Place and Identity: Commemoration and Remembrance of War and Conflict,* Oxford and New York: Routledge, 1–16.

Dwyer, O. and Alderman, D. (2008) 'Memorial landscapes: Analytic questions and metaphors', *GeoJournal,* 73(3): 165–178.

Fenton, D. (2013) *New Zealand and the First World War, 1914–1919,* Auckland: Penguin.

Foster, J. (2004) 'Creating a temenos, positing "South Africanism": Material memory, landscape practice and the circulation of identity at Delville Wood', *Cultural Geographies,* 11: 259–290.

Frame, T. (ed.) (2016) *Anzac Day: Then and Now,* Sydney: NewSouth Publishing.

Gegner, M. and Ziino, B. (2012) 'Introduction: The heritage of war: Agency, contingency, identity', in M. Gegner and B. Ziino (eds.), *The Heritage of War,* Oxford and New York: Routledge, 1–15.

Gordon, D. and Osborne, B. (2004) 'Constructing national identity in Canada's capital, 1900–2000: Confederation square and the National War Memorial', *Journal of Historical Geography,* 30(4): 618–642.

Gough, P. (2004) 'Sites in the imagination: The Beaumont Hamel Newfoundland Memorial on the Somme', *Cultural Geographies,* 11: 235–258.

Gough, P. and Morgan, S. (2004) 'Manipulating the metonymic: The politics of civic identity and the Bristol cenotaph, 1919–1932', *Journal of Historical Geography,* 30(4): 665–684.

Graham, B. (1994) 'No place of the mind: Contested protestant representations of Ulster', *Ecumene,* 1(3): 257–281.

Graham, B., Ashworth, G. and Tunbridge, J. (2000) *A Geography of Heritage (Power, Culture and Economy),* London: Arnold Hodder Headline.

Graham, B. and Shirlow, P. (2002) 'The battle of the Somme in Ulster memory and identity', *Political Geography,* 21: 881–904.

Grayson, R. and McGarry, F. (eds.) (2016) *Remembering 1916: The Easter Rising, the Somme and the Politics of Memory in Ireland,* Cambridge: Cambridge University Press.

Harvey, D.C. (2017) 'Critical heritage debates and the commemoration of the First World War: Productive nostalgia and discourses of respectful reverence during the centenary', in H. Silverman, E. Waterton and S. Watson (eds.), *Heritage in Action: Making the Past in the Present,* Cham, Switzerland: Springer, 107–120.

Heffernan, M. (1995) 'For ever England: The Western front and the politics of remembrance in Britain', *Ecumene,* 2(3): 293–323.

Hoelscher, S. and Alderman, D. (2004) 'Memory and place: Geographies of a critical relationship', *Social & Cultural Geography,* 5(3): 347–355.

Hoffenberg, P. (2001) 'Landscape, memory and the Australian war experience, 1915–18', *Journal of Contemporary History,* 36(1): 111–131.

Holbrook, C. and Ziino, B. (2015) 'Family history and the Great War in Australia', in B. Ziino (ed.), *Remembering the First World War*, Oxford and New York: Routledge, 39–55.

Inglis, K. (2005) *Sacred Places: War Memorials in the Australian Landscape*, Melbourne: Melbourne University Press.

Jarman, N. (1999) 'Commemorating 1916, celebrating difference: Parading and painting in Belfast', in A. Forty and S. Kuchler (eds.), *The Art of Forgetting*, Oxford: Berg, 171–195.

Johnson, N. (1999) 'The spectacle of memory: Ireland's remembrance of the Great War, 1919', *Journal of Historical Geography*, 25(1): 36–56.

Johnson, N. (2000) 'Historical geographies of the present', in B. Graham and C. Nash (eds.), *Modern Historical Geographies*, Harlow: Pearson Education Limited, 251–272.

Johnson, N. (2003) *Ireland, the Great War and the Geography of Remembrance*, Cambridge: Cambridge University Press.

Jones, H. (2014) 'The Great War: How 1914–18 changed the relationship between war and civilians', *The RUSI Journal*, 159(4): 84–91.

Killingray, D. (2012) 'The war in Africa', in J. Horne (ed.), *A Companion of World War I*, Chichester: Wiley-Blackwell, 112–126.

King, A. (1998) *Memorials of the Great War in Britain: The Symbolism and Politics of Remembrance*, Oxford and New York: Berg.

King, A. (1999) 'Remembering and forgetting in the public memorials of the Great War', in A. Forty and S. Kuchler (eds.), *The Art of Forgetting*, Oxford: Berg, 147–169.

King, A. (2010) 'The Afghan War and "postmodern" memory: Commemoration and the Dead of Helmand', *The British Journal of Sociology*, 61(1): 1–25.

Legg, S. (2007) 'Reviewing geographies of memory/forgetting', *Environment & Planning A*, 39(2): 456–466.

Macleod, J. (2013) 'Britishness and commemoration: National memorials to the First World War in Britain and Ireland', *Journal of Contemporary History*, 48(4): 647–665.

Maguire, A. (2016) 'Looking for "home"? New Zealand soldiers visiting London during the First World War', *The London Journal*, 41(3): 281–298.

McAuley, J. (2014) 'Divergent memories: Remembering and forgetting the Great War in loyalist and nationalist Ireland', in S. Sumartojo and B. Wellings (eds.), *Nation, Memory and Great War Commemoration: Mobilizing the Past in Europe, Australia and New Zealand*, Bern: Peter Lang Academic Publishers, 119–132.

McCarthy, M (2005) 'Historico-geographical explorations of Ireland's heritage: Towards a critical understanding of the nature of memory and identity', in M. McCarthy (ed.), *Ireland's Heritage: Critical Perspectives on Memory and Identity*, Aldershot: Ashgate Publishing Limited, 3–51.

McCarthy, M. (2013) *Ireland's 1916 Rising: Explorations of History-Making, Commemoration and Heritage in Modern Times*, Farnham: Ashgate, 2013.

McGeachen, C. (2014) 'Historical geography I: What remains?', *Progress in Human Geography*, 38(6): 824–837.

Mckenna, M. (2014) 'Keeping in step: The Anzac "resurgence" and "military heritage" in Australia and New Zealand', in S. Sumartojo and B. Wellings (eds.), *Nation, Memory and Great War Commemoration: Mobilizing the Past in Europe, Australia and New Zealand*, Bern: Peter Lang Academic Publishers, 151–167.

Meyer, J. (2009) *Men of War: Masculinity and the First World War in Britain*, Basingstoke: Palgrave Macmillan.

Miles, S. (2016) *The Western Front: Landscape, Tourism and Heritage*, Barnsley: Pen & Sword Archaeology.

Moriarty, C. (1995) 'The absent dead and figurative First World War memorials', *Transactions of the Ancient Monuments Society*, 39: 7–40.

Moriarty, C. (1997) 'Private grief and public remembrance: British First World War memorials', in M. Evans and K. Lunn (eds.), *War and Memory in the Twentieth Century*, Oxford and New York: Berg, 125–142.

Moriarty, C. (1999) 'The material culture of Great War remembrance', *Journal of Contemporary History*, 34(4): 653–662.

Morris, M. (1997) 'Gardens "For Ever England": Landscape, identity and the First World War British cemeteries of the Western Front', *Ecumene*, 4: 410–434.

Morrissey, J. (2005) 'A lost heritage: The Connaught Rangers and multivocal Irishness', in M. McCarthy (ed.), *Ireland's Heritage: Critical Perspectives on Memory and Identity*, Aldershot: Ashgate Publishing Limited, 71–87.

Mosse, G. (1990) *Fallen Soldiers: Reshaping the Memory of the World Wars*, Oxford: Oxford University Press.

Muzaini, H. and Yeoh, B. (2007) 'Memory-making "from below": Rescaling remembrance at the Kranji War Memorial and Cemetery, Singapore', *Environment & Planning A*, 39(6): 1288–1305.

Muzaini, H. and Yeoh, B. (2016) *Contested Memoryscapes: The Politics of Second World War Commemoration in Singapore*, Oxford and New York: Routledge.

Mycock, A. (2014a) 'The First World War Centenary in the UK: "A truly national commemoration"?', *The Round Table: The Commonwealth Journal of International Affairs*, 103(2): 153–163.

Mycock, A. (2014b) 'The politics of the Great War Centenary in the United Kingdom', in S. Sumartojo and B. Wellings (eds.), *Nation, Memory and Great War Commemoration: Mobilizing the Past in Europe, Australia and New Zealand*, Bern: Peter Lang Academic Publishers, 99–118.

Mycock, A., Sumartojo, S. and Wellings, B. (2014) '"The centenary to end all centenaries": The Great War, nation and commemoration', in S. Sumartojo and B. Wellings (eds.), *Nation, Memory and Great War Commemoration: Mobilizing the Past in Europe, Australia and New Zealand*, Bern: Peter Lang Academic Publishers, 1–24.

Olusoga, D. (2014) *The World's War: Forgotten Soldiers of Empire*, London: Head of Zeus Ltd.

Parker, P. (2009) *The Last Veteran: Harry Patch and the Legacy of War*, London: Fourth Estate, HarperCollins.

Pennell, C. (2012a) *A Kingdom United: Popular Responses to the Outbreak of the First World War in Britain and Ireland*, Oxford: Oxford University Press.

Pennell, C. (2012b) 'Popular history and myth-making: The role and responsibility of First World War historians in the centenary commemorations, 2014–2018', *Historically Speaking*, 13(5): 11–14.

Reeves, K., Bird, G., Stichelbaut, B. and Bourgeois, J. (eds.) (2016) *Battlefield Events: Landscape, Commemoration and Heritage*, Oxford and New York: Routledge.

Reynolds, D. (2013) The Long Shadow: The Great War and the Twentieth Century, London: Simon & Schuster.

Rowlands, M. (1999) 'Remembering to forget: Sublimation as sacrifice in war memorials', in A. Forty and S. Kuchler (eds.), *The Art of Forgetting*, Oxford: Berg, 129–145.

Saunders, N. (2004) 'Material culture and conflict: The Great War 1914–2003', in N. Saunders (ed.), *Matters of Conflict: Material Culture, Memory and the First World War*, Oxford: Routledge, 5–25.

Saunders, N. and Cornish, P. (eds.) (2009) *Contested Objects: Material Memories of the Great War*, London and New York: Routledge.

Scates, B. (2006) *Return to Gallipoli: Walking the Battlefields of the Great War*, Cambridge: Cambridge University Press.

Scates, B. (2009) 'Manufacturing memory at Gallipoli', in M. Keren and H. Herwig (eds.), *War Memory and Popular Culture: Essays on Modes of Remembrance and Commemoration*, Jefferson, NC and London: McFarland & Company Inc. Publishers, 57–75.

Scates, B. (2016) 'The Unquiet Grave: Exhuming and reburying the dead of Fromelles', in K. Reeves, G. Bird, L. James, B. Stichelbaut and J. Bourgeois (eds.), *Battlefield Events: Landscape, Commemoration and Heritage*, Oxford and New York: Routledge, 13–27.

Sheftall, M. (2009) *Altered Memories of the Great War: Divergent Narratives of Britain, Australia, New Zealand and Canada*, London and New York: I.B. Tauris & Co Ltd.

Slade, P. (2003) 'Gallipoli thanatourism: The meaning of ANZAC', *Annals of Tourism Research*, 30: 779–794.

Smith, R. (2004) *Jamaican Volunteers in the First World War: Race, Masculinity and the Development of National Consciousness*, Manchester: Manchester University Press.

Sørensen, M. and Viejo-Rose, D. (2015) 'Introduction: The impact of conflict on cultural heritage: A biographical lens', in M. Sørensen and D. Viejo-Rose (eds.), *War and Cultural Heritage: Biographies of Place*, New York: Cambridge University Press, 1–17.

Stephens, J. (2007) 'Memory, commemoration and the meaning of a Suburban War memorial', *Journal of Material Culture*, 12(3): 241–261.

Stevenson, D. (2005) *1914–1918, the History of the First World War*, London: Penguin Books.

Strachan, H. (2003) *The First World War: A New History*, London: Simon & Schuster.

Sumartojo, S. (2015) '"You aren't an Aussie if you don't come": National Identity and Visitors' Practices at the Australian National Memorial, Villers-Bretonneux', in J. Lossau and Q. Steven (eds.), *The Use of Art in Public Space*, London: Routledge, 131–144.

Sumartojo, S. (2016a) 'Commemorative atmospheres: Memorial sites, collective events and the experience of national identity', *Transactions of the Institute of British Geographers*, 41(4): 541–553.

Sumartojo, S. (2016b) 'Anzac atmospheres', in D. Drozdzewski, S. De Nardi and E. Waterton (eds.), *Memory, Place and Identity: Commemoration and Remembrance of War and Conflict*, Oxford and New York: Routledge, 189–204.

Sumartojo, S. and Wellings, B. (eds.) (2014) *Nation, Memory and Great War Commemoration: Mobilizing the Past in Europe, Australia and New Zealand*, Bern: Peter Lang Academic Publishers.

Sumartojo, S. and Wellings, B. (eds.) (2017) *Commemorating Race and Empire in the Great War Centenary*, Liverpool and Marseille: Liverpool University Press/Presses Universitaires de Provence.

Switzer, C. and Graham, B. (2010) 'Ulster's love in letter'd gold': The Battle of the Somme and the Ulster memorial tower, 1918–1935', *Journal of Historical Geography*, 36(2): 183–193.

Todman, D. (2005) *The Great War: Myth and Memory*, London: Hambledon Continuum.

Todman, D. (2008) 'The First World War in contemporary British popular culture', in H. Jones, J. O'Brien and C. Schmidt-Supprian (eds.), *Untold War: New Perspectives in First World War Studies*, Boston and Leiden: Brill, 417–441.

Todman, D. (2009) 'The ninetieth anniversary of the Battle of the Somme', in M. Keren and H. Herwig (eds.), *War Memory and Popular Culture: Essays on Modes of Remembrance and Commemoration*, Jefferson, NC and London: McFarland & Company Inc. Publishers, 23–40.

Vance, J. (1997) *Death So Noble: Memory, Meaning and the First World War*, Vancouver: University of British Columbia Press.

Wallis, J. (2015) 'Great-grandfather, what did you do in the Great War?: The phenomenon of conducting First World War family history research', in B. Ziino (ed.), *Remembering the First World War*, Oxford and New York: Routledge, 21–38.

Walsh, M. and Varnava, A. (eds.) (2017) *The Great War and the British Empire: Culture and Society*, Oxford and New York: Routledge.

Watson, A. (2015) *Ring of Steel: Germany and Austria-Hungary at War, 1914–1918*, London: Penguin Books.

Wellings, B. (2014) 'Lest you forget: Memory and Australian nationalism in a Global Era', in S. Sumartojo and B. Wellings (eds.), *Nation, Memory and Great War Commemoration: Mobilizing the Past in Europe, Australia and New Zealand*, Bern: Peter Lang Academic Publishers, 45–59.

West, B. (ed.) (2016) *War Memory and Commemoration*, Oxford and New York: Routledge.

Wilson, R. (2007) 'Archaeology on the Western Front: The archaeology of popular myths', *Public Archaeology*, 6(4): 227–241.

Wilson, R. (2011) '"Tommifying" the Western Front, 1914–1918', *Journal of Historical Geography*, 37(3): 338–347.

Wilson, R. (2013) *Cultural Heritage of the Great War in Britain*, Farnham: Ashgate.

Winter, J. (1999) 'Forms of Kinship and remembrance in the aftermath of the Great War', in J. Winter and E. Sivan (eds.), *War and Remembrance in the Twentieth Century*, Cambridge: Cambridge University Press, 40–60.

Winter, J. (2006) *Remembering War: The Great War between Memory and History in the Twentieth Century*, New York: Yale University Press.

Winter, J. (ed.) (2014) *The Cambridge History of the First World War (Volumes I-III)*, Cambridge and New York: Cambridge University Press.

Ziino, B. (2012) '"We are talking about Gallipoli after all": Contested narratives, contested ownership and the Gallipoli Peninsula', in M. Gegner and B. Ziino (eds.), *The Heritage of War*, Oxford and New York: Routledge, 142–159.

Ziino, B (2015) 'Introduction: Remembering the First World War today', in B. Ziino (ed.), *Remembering the First World War*, Oxford and New York: Routledge, 1–17.

Part 1

Rethinking, and looking beyond the front line

2   Congested terrain

Contested memories. Visualising the multiple spaces of war and remembrance

Paul Gough

'Stasis' is widely accepted as the pre-eminent condition of the conflict on the Western Front; a war of congealment, fixity and stagnant immobility fought from defensive earthworks that were intended to be temporary but quickly became permanent.

In the battle zones, a new spatial order emerged. Beyond the superficial safety of the front-line parapet was No-Man's-Land – a liminal, unknown space, a 'debatable land' that could not be fully owned or controlled. Far beyond lay a green and unspoilt distance, a 'Promised Land' that was forever locked in an unattainable future. This was the domain of imperial development and potential exploitation.

This chapter explores the spatiality of conflicts on the Great War battlefield and draws on the work of several British artists, cartographers and surveyors who attempted to explore and lend visual form to the chaos. Through the act of mapping and drawing they attempted to systematize the outward devastation, whereby trees would become datum points, emptiness was labelled, and the few fixed features of the ravaged land became the immutable coordinates of a functional terrain, a strategic field, where maps where predicated as much on time as of place.

Part one: drawing in dystopia

On 13 November 1917, after several weeks of drawing in the Ypres Salient, the young British painter Paul Nash wrote to his wife describing what he had witnessed:

> The rain drives on, the stinking mud becomes more evilly yellow, the shell holes fill up with green-white water, the roads and tracks are covered in inches of slime, the black dying trees ooze and sweat and the shells never cease. They alone plunge overhead, tearing away the rotting tree stumps, breaking the plank roads, striking down horses and mules, annihilating, maiming, maddening, they plunge into the grave that is this land; one huge grave, and cast up on it the poor dead. It is unspeakable, godless, hopeless.
>
> (Nash 1948: 186)

From this extraordinary vision, Nash, a seconded infantry officer with recent combat experience, drew a powerful conclusion:

> I am no longer an artist interested and curious, I am a messenger who will bring back word from the men who are fighting to those who want the war to go on for ever. Feeble, inarticulate, will be my message, but it will have a bitter truth, and may it burn their lousy souls.
>
> (Nash 1948: 186)

Of the hundreds of official war artists who strived to describe the dystopian appearance of the Western Front, few wrote with such a scalding anger as Nash. Emboldened by his own front-line experience as an infantry officer he channelled his emotional intensity into a suite of artworks inspired by the muddy wastes and hollowed slopes of the Salient. Along with a handful of young Modernist artists he created a new calligraphy of war; drawings scored and scratched with harsh diagonals; incessant rainfall engraved in stabbing lines across the surface; the ashen wastelands captured in dense strokes of the pen. Nothing he saw or felt daunted him: neither the weird sight of a tree-trunk adorned in barbed wire, a night-time barrage on the sodden battlefield, or a close-up of driving raindrops falling heavily into the convulsed earth. His was a radical vision ideally matched to this 'phantasmagoric world' (Eates 1973: 22).

Like many of his fellow young Modernist painters, Nash embraced the diagonal energy and strict tonalities of his Vorticist contemporaries. Armoured by the abrupt angularities and jagged idiom of their practice Nash and other painters, confronted by the war in Europe abandoned their pastoral visions, replacing them with splintered woods, violated nature, wasted panoramas. The shock was laced with ironies. In one letter, Nash recalled walking through a wood (or at least what remained of it after heavy shelling) when it was little more than 'a place with an evil name, pitted and pocked with shells, the trees torn to shreds, often reeking with poison gas' (Nash 1948: 187). A few days later, to his astonishment, that 'most desolate ruinous place' was drastically changed. It was now 'a vivid green', bristling with buds and fresh leaf growth:

> The most broken trees even had sprouted somewhere and in the midst, from the depth of the wood's bruised heart poured out the throbbing song of a nightingale. Ridiculous mad incongruity! One can't think which is the more absurd, the War or Nature.
>
> (Nash 1948: 187)

Critics were astonished at his daring. His new work, wrote one, heralded 'an actuality, an immediacy, that brought to life everything about the front which people had read and heard, but had found themselves quite unable to visualize' (cited in Eates 1973: 22). Few contemporary British artists were as sensitive to the despoliation of the natural order. Nash was both spellbound and aghast at the sight of splintered copses and dismembered trees, seeing in their shattered limbs an equivalent to the human carnage that lay all around or even hung in shreds from the eviscerated treetops. In new batches of work the trees remain inert and gaunt, failing to respond to the shafts of sunlight; their branches dangle lifelessly 'like melancholy tresses of hair', mourning the death of the world and its values that Nash held so dear. Writing in the foreword to the *Void of War* exhibition in London in May 1918, the writer and director of Propaganda in France, Arnold Bennett caught its essential spirit:

> Lieutenant Nash has seen the Front simply and largely. He has found the essentials of it – that is to say, disfigurement, danger, desolation, ruin, chaos – the little figures of men creeping devotedly and tragically over the waste. The convention he uses is ruthlessly selective. The wave-like formations of shell-holes, the curve of shell-bursts, the straight lines and sharply defined angles of wooden causeways, decapitated trees, the fangs of obdurate masonry, the weight of heavy skies, the human pawns of battle.
>
> (Bennett 1918: 3)

In developing a new syntax of despoliation, Nash's work was attuned to the radically new conditions of modern warfare. Like fellow front-line artist C.W.R. Nevinson, he grasped the industrial scale of the war machine and the vast shapelessness of the battlefield. His calligraphic dexterity and enriched imagination made sense of the weird incongruities of the Western Front battlescape; its outward inertia and the constant sense of menace and foreboding (Gough 2010).

To even the most observant eye the landscape seemed outwardly deserted but the dead lay just beneath its ruptured surface and the living led an ordered and disciplined (although often very perilous) existence in underground shelters and deep chambers. It was one of the greatest contradictions of modern warfare: here was a landscape that gave the appearance by daylight of being empty but was emphatically not. It teemed with invisible life. Furthermore, it was always being scrutinised. Mechanical eyes far above the pockmarked ground patiently gridded and re-gridded the lunar surface, systematically mapping what appeared to be an unmappable surface. The blasted topography was fixed, measured, calibrated and then named. It was not so much a landscape, as an uncanny 'paradox of measurable nothingness' (Weir 2007: 43).

Yet, in considering the lunar face of No Man's Land the writer Reginald Farrer (1918) suggested that it was misleading to regard the 'huge, haunted solitude' of the modern battlefield as empty. 'It is more', he argued, 'full of emptiness . . . an emptiness that is not really empty at all' (Farrer 1918: 113). Nash adapted Farrer's conception of the 'Void of War' to re-determine the inverted spatiality of the modern battlefield, populating its emptiness with latent violence. He knew from his time in the infantry that on the modern battlefield danger was omnidirectional, threat lay in every conceivable direction not merely from the fixed enemy line to one's front but from underneath, overhead, and from behind. This need to create a new, revised omni-spatial awareness may explain why photography so failed to convey the new de-materialities and vast chaos of war (Kern 1983: 147).

Not far from where Nash had hunched in concentration over his drawing board, the official Australian photographer Frank Hurley adjusted the cumbersome paraphernalia of his plate-glass camera and tripod and staring nonplussed into the deserted waste of the Ypres Salient. A veteran of Ernest Shackleton's ill-fated second expedition to the Antarctic in 1914–16, he was trying to capture the sprawling mess of the battlefield in a single frame:

> Everything is on such a wide scale. Figures scattered, atmosphere dense with haze and smoke – shells that would simply not burst when required. All the elements of a picture were there, could they but be brought together and condensed.
>
> (Bickel 1980: 61)

Having endured the barren vistas of the southern ice fields, Hurley was deeply frustrated by the diffuse character of the war in Flanders – the lack of focal points, its vastness and sprawling anonymity. Other equally perplexed official photographers – such as the Canadian, Ivor Castle – fabricated their own battle compositions in the darkroom, combining negatives one on top of another to create a composite version of trench warfare. With a little judicious cropping he discovered that an innocent image could be transformed into something more martial, even sensational, and, like Hurley, he courted controversy by superimposing shrapnel bursts and fighter aeroplanes into the clear skies. Flat and level for long distances, the Flanders landscape was a convenient and uncomplicated backdrop for these carefully contrived multi-layered 'combats' (Carmichael 1989: 50–55).

But photography could not visualise emptiness; it could only allude to absences. Even words failed to convey the intensity of its emptiness. Faced with the lunar features of the Western Front, its paralysing horror (Wylie 2007), the imagination froze:

> It seemed quite unthinkable that there was another trench over there a few yards away just like our own . . . Not even the shells made that brooding watchfulness more easy to grasp; they only made it more grotesque. For everything was so paralysed in calm, so unnaturally innocent and bland and balmy. You simply could not take it in.
>
> (Farrer 1918: 113)

Part two: 'this tree sense'

Having explored the spectral geographies of the Western Front as experienced visually and viscerally by a number of artists and photographers, we turn next to the datum points of that traumatised landscape, in particular the natural vegetation, woods, and trees that were to become key components in the conflict's geographical narrative.

Across the historically verdurous plains of Flanders, Artois and Picardy, trees offered vantage and protection. They provided raw materials and nourishment. They lay in thick forests as well as compact copses. Overwhelmed by the maelstrom of war their relentless decimation was remembered by many who witnessed it:

> I never lost this tree sense: to me half the war is a memory of trees; fallen and tortured trees; trees untouched in summer moonlight, torn and shattered winter trees, trees green and brown, grey and white, living and dead. They gave their names to roads and trenches, strong points and areas. Beneath their branches I found the best and the worst of war.
>
> (Talbot Kelly 1980: 5)

Amidst the devastation, as Saunders (2014) has pointed out, human relationships with nature, and with trees especially, were forced to change. Wood, leaves, roots and branches took on new symbolic meanings and sensorial qualities. Camouflage mimicked their clever patterns, fake trees concealed snipers and observers, copses hid entire batteries of artillery, subterranean dugouts were filled with the smell of freshly cut wood, and ancient willows were bent into new shapes as the revetments for front-line trenches. Woodlands – Mametz Wood on the Somme; Polygon Wood on the Ypres Salient, for example, were turned into strongholds fiercely fought over by both sides. Singular, isolated trees became a registration point for enemy artillery. Soldiers soon learned to avoid them:

> The 'Lone Tree' was the assembly point for the wounded, and all around on the grass there were dozens of wounded on stretchers waiting to be taken down by the ambulance column. This tree was a favourite for the German artillery and I could never understand why the wounded, transport, cookers and ambulances were allowed to congregate in this area. Apparently, somebody later recognised the danger and the tree was felled.
>
> (Stuart Dolden 1980: 39)

In a war seeped with irony, was this not one of the most cruel mockeries of that terrible war? A once attractive, proud (in the sense of free-standing) tree was now to be feared, and regarded as a point of maximum danger. What was once an icon of nature that might be cherished as a place of refuge and shade, had become wickedly inverted by the chaos of war. During the course of the conflict, trees, and especially small woods, were to become notorious traps. Mansel Copse, Inverness Wood, Thiepval Forest and dozens of others, some no larger than a hockey pitch, became infamous killing grounds, sites of extraordinary mayhem contested by rival troops for months (Gough 2004: 237–238). On trench maps, the very words – 'copse', 'wood', even 'forest' – soon became irrelevant as trees were felled by artillery shells, reduced to splinters, charred by fire, and felled for military use. So systematic was the destruction and so ruthless the cutting it was estimated that it would take half a century for many areas to be able to produce decent timber again. Indeed, French horticulturalists coined the term 'forest trauma' (Clout 1996: 30–34). Today a few of these 'ancient' trees remain:

a hornbeam that miraculously survived the total devastation of Delville Wood, its trunk still stippled with metallic fragments of ordnance; another rooted in a barrel of cement marks the point in no-mans-land where hundreds of soldiers from the Newfoundland Regiment collected in search of a gathering-point, only to meet a sudden, bloody end in the first 40 minutes of the first day of the Battle of the Somme. It stands today – the 'Danger Tree' – an ashen stalk once promising succour, but delivering only pain and death (Cave 1994).

How could painters such as Nash and Nevinson, and their many contemporaries, relate to the shattered mayhem that constituted the flattened landscape of the scorched battlegrounds? There were precedents in romantic painting, gothic literature, and in turn-of-the-century renditions of apocalyptic cityscapes. To augment these precedents there is another teasing question. Could Nash, Nevinson or any other front-line soldier-artist have been aware that a century earlier, a young British painter had devised an extraordinary image of an equally poisoned and murderous landscape? At its centre was the notorious 'Upas Tree'.

> This fabulous tree was said to grow on the island of Java, in the midst of a desert formed by its own pestiferous exhalations. These destroyed all vegetable life in the immediate neighbourhood of the tree, and all animal life that approached it. Its poison was considered precious, and was to be obtained by piercing the bark, when it flowed forth from the wound. So hopeless, however, and so perilous was the endeavour to obtain it, that only criminals sentenced to death could be induced to make the attempt, and as numbers of them perished, the place became a valley of the shadow of death, a charnel-field of bones.
>
> (Redgrave 1886: 483)

The fable of the dreaded *Upas Tree* is based on the tale of the poisonous anchar tree, first revealed by the eighteenth-century botanist Erasmus Darwin. During the Romantic era, it became a familiar and potent image adopted by such poets as Samuel Taylor Coleridge, occurring in Lord Byron's *Childe Harolde's Pilgrimage* and in Robert Southey's epic poem *Thalaba*. But probably the most significant evocation of this bizarre plant is in the vast canvas painted by the Bristol-based painter Francis Danby. He painted *The Upas, or Poison* Tree, *in the Island of Java* in 1819, when he was just 26; a year later, it heralded his triumphal arrival on the London art scene. By the standards of its time, it is not a very large painting – it is some 5 feet by 7 – but its sense of scale is striking. A cowering figure is dwarfed by the landscape; the tree is little but an upright bony stalk but dominates the surrounding terrain. The rocky valley – said to have been modelled on the Avon Gorge – is a hazardous, leafless place of crags and fissures, surrounded by the corpses of fated adventurers (Greenacre 1988: 89–91).

The painting was equally as doomed. After mixed reviews, it sold for £150 but the fee went straight to a host of creditors, Danby having worked up a large debt during his stay in Bristol. Within years, the picture had declined in quality: sloppy technique and thick varnish had rendered the image almost unreadable. By 1857, the picture was barely visible: it was as if the poisonous exhalations of the motif had spread to the very paint surface. Extensive cleaning and removal of layers of dark varnish have revived the painting a little, but even after laborious conservation it still catches the light and is difficult to see. Like the eponymous tree, one approaches the large canvas with squinting eyes and a measured tread. The tale of the Upas Tree may seem somewhat tangential to this examination of the geographies of war on the Western Front. Yet the icon of 'Upas' activates the spectral turn identified by Sebald (1995). Transposed to a battlefield setting, it immediately conjures the relevance of the ghostly (Wylie 2007). The worlds Sebald describes, and Wylie analyses, are those occupied by the displaced, traumatised and exiled. These are the very same worlds that confronted artists, writers, geographers during the conflict and in the years after the Great War. Haunted by harrowing experiences they moved through a dystopia cleared of occupants through expulsion and exclusion, saturated with traumatic memory, and rendered nondescript by the impact of sustained static warfare. With its 'sudden violences and long stillnesses' it remained for many a 'place of enchantment' (Jones 1937: *x*).

Ninety-nine years after Danby finished his painting, war artist Paul Nash, also in his late twenties, was struggling with his own '*magnum opus*', the huge canvas now known as *The Menin Road*. This was actually his first foray into oil paint, an act – as he put it – of 'great audacity' (Abbot and Bertram 1955: 98). Like Danby's image, it describes a blighted land; the dystopian wilderness of the Western Front – a pestilent waste of shattered trees, toxic soils and scattered bones. Perhaps here, for the first time since Danby depicted the corrupted 'poison anchar of Java', might be found the *Upas* in its modern spectral incarnation.

In fact, Nash served for only a very short time on the front line. In May 1917, after less than eight weeks in Flanders, he was injured in a bad fall and invalided home days before many of his fellow officers were killed in an offensive. However, what he saw in that short time had a radical impact on him and his art. He was most stunned at the perseverance of nature (Nash 1948: 186). Although war had wreaked its havoc, he observed that nature was extraordinarily resilient. Nash was both bemused and maddened by the strange absurdities all around him, unsure whether to aim his eloquent anger at the war, at nature, or at 'we poor beings [who] are double enthralled' (Nash 1948: 186). The war radicalised his art. Gone were the rather sentimental and derivative ideas borrowed from his Pre-Raphaelite phase, sidelined was the chivalric idealism of his late youth, to be replaced by a tougher language that matched the novel and grim conditions he had witnessed. Never before had he been subject to circumstances and places that were so pitiless, cruel, and malignant. For possibly the first time as a painter, Nash was seeing for *himself*, not applying the tired conventions of an art-practice or imposing the vision of others.

Although Nash and Nevinson bucked artistic convention to create a new iconography of art and practice, we should turn now to other artists who had to actually embrace convention for very different ends.

Part three: 'fields of vision'

Throughout the war artistically talented soldiers of all ranks in the British Army, especially those who had some grasp of landscape painting, found themselves sought out to work in the Camouflage Corps or for the Field Survey. Not all went willingly. Painter Harry Bateman, having volunteered for overseas service with the Royal Field Artillery, ignored a sergeant's request at their first parade for any artist present to make himself known. He 'remained silent as he wanted to go and fight' (Brown 1978: 185; Gough 2009: 240). Others found that their skills were deemed inappropriate: the painter and poet David Jones, serving with the 15th (London Welsh) Battalion, Royal Welch Fusiliers, had five years art school training to his credit when he was recommended to the 2nd Field Survey Company based at Second Army Headquarters, Cassel. Jones, however, lacked the technical skills required for map drawing and was instead sent to one of the Company's four observation groups as a Survey Post observer. Having already been promoted sideways Jones did not last much longer as an observer: 'Got the sack from that job because of my inefficiency in getting the right degrees of enemy gunflashes' (Hague 1980: 241; Chasseaud 2013: 19). Another artist, a contemporary of Paul Nash, failed abjectly in the simple military task of 'breaking ground':

> From the OP [Observation Post] I saw a completely featureless landscape, save here and there a few broken sticks of trees. I made a pencil drawing of this barren piece of ground, but what use my superiors would be able to make of this sketch I could not imagine.
>
> (Roberts 1974: 27–28)

Thus ended the modernist painter William Roberts' first and only foray into reconnaissance drawing. In fact, tellingly few of the other young 'moderns' serving in the armed forces during the Great War could channel or discipline their artistic tendencies in the pursuit of technical objectivity. Although they could render the dystopian face of the war and its impact on nature, they could not master the simplified graphic language required of the military sketch. Yet there were others who advertised their skills quite freely. Adrian Hill, one of the youngest soldier-artists to eventually work for the official government war art schemes, combined his drawing abilities with his work in a Scouting and Sniping Section of the Honorable Artillery Company (Gough 2010). After the war, he recalled a typical patrol into No Man's Land:

> I advanced in short rushes, mostly on my hands and knees with my sketching kit dangling round my neck. As I slowly approached, the wood gradually took a more definite shape, and as I crept nearer I saw that what was hidden from our own line, now revealed itself as a cunningly contrived observation post in one of the battered trees.
>
> (Hill 1930: 16)

Drawing for military purposes can in fact be separated into two distinct fields of vision. These correspond approximately to the different arms of the military: on the one hand are those drawings made during mobile reconnaissance, usually by light cavalry patrols or small units of advanced infantry. These are used to record intelligence about enemy positions and key terrain. On the other hand, there are drawings known as panoramas that are made from a static position, usually an elevated vantage point that commands an uninterrupted view of the enemy front. These are normally drawn by specially trained artillery or engineer officers and are vital for indicating targets and determining range and arc of medium- and long-range artillery fire.

The principle of panoramic drawing, when used in an artillery sense, developed from the role of the Forward Observation Officer (FOO) whose task was to direct the fire of guns located much further back from his post on the edge of known and secure ground. Through close scrutiny of enemy-held territory, the FOO was able to engage targets very rapidly across the whole arc of view. If a number of targets had already been pre-registered, and engaged to an exact point on the ground, that point could be marked on a drawn panorama. This drawing would also be copied to the gunners in the rear who would then be able to engage the same target number with greater speed and efficiency. In effect, the panorama became a surrogate view for the distant artillery blinded by dead ground or topographic barriers.

Whereas the patrol sketch invariably takes the form of a collage of hasty impressions later re-arranged to create a tactical narrative, the panorama is essentially concerned with scopic control and spatial dominance. The artillery panorama works on the same premise as military mapping; if performed diligently and with a high attention to accuracy detailed surveillance and graphic survey will eventually neutralise a dangerous terrain and assure mastery over it (Alfrey and Daniels 1990). In a similar spirit, Foucault (1975) wrote of the system of permanent registration that operated in the plague town in the seventeenth century. On the septic terrain of the First World War battlefield, the panoramic drawing was an integral part in segmenting and immobilizing perceived space. The stasis of the battle line, however, meant that the panoptic ideal could never be attained: dead ground (space beyond or concealed from retinal view), and camouflage (employed to aid deception and concealment) were constant and deliberate frustrations to retinal surveillance. Foucault's concept of a transparent space was constantly frustrated by the fissured and volatile landscape of the battlefield. The military sketch, however, was the nearest graphic equivalent of Bentham's paradigm: it provided systematic observation 'in which the slightest movements are supervised, in which all events are recorded' (Foucault 1975: 197).

The result of this obsession with registering, recording and calibration was the mass production of surveys, maps, and drawings, produced constantly by the opposing armies, and updated daily by fresh photographs taken from the 'eye in the sky'. The scale of endeavor was extraordinary. Carmichael (1989) reveals in her research into military aerial photography that the Royal Flying Corps (later Royal Air Force) produced an updated photographic record of the entire British section of the Western Front every day. Processed, printed, and displayed for intelligence purposes these photographs were augmented by the military sketches and panoramas drawn from ground level. As the primary sources of geospatial intelligence these observations and photographs states Saint-Amour (2003) in his appraisal of modernist reconnaissance, 'were projected onto the geometric order of the map, which was animated by the mechanical cadence of the military time-table' (Saint-Amour 2003: 353).

Indexical in its thoroughness and relentless in its pursuit of accuracy this process of objectification mirrored the complex bureaucracies developed by the industrial armies during total war. The habit persisted after the war, at least within the bureaucracy of the victors. The administration of death echoed the same military machine that had become rationalised, routinised, and standardised during five years of conflict.

Furthermore, after the Armistice, the dead, as Heffernan (1995: 113) points out, were not allowed 'to pass unnoticed back into the private world of their families'. They became 'official property' to be accorded appropriate civic commemoration in solemn monuments of official remembrance. Having seized the ideological authority over the rights of the individual citizen, and because large tracts of foreign territory were deemed now to be possessed by its dead, the British Empire, through its constituent representatives, negotiated schemes to enclose portions of land in perpetuity. During this process, the fusion of body and territory became essential to the understanding of geographical narratives. As Malvern (2001: 57) reminds us, the trope of land- and landscape-as-body was a constant evocation during and immediately after the war. Contemporary war artists understood this meshing of forms; it helped fuel their imagination and resulted in some of the most searing and abstracted imagery of the conflict. Military sketching by comparison sought to isolate and dismiss such abstract subjectivity, eschewing artistic impulse in the pursuit of immutable fact and verifiable indicators.

Part four: malign space and bushy topped trees

Military sketching – both for short-view use and for long-distance panoramas – had to make visual sense of a battleground that had in many places been bitterly contested for many years. To borrow Gregory's memorable term (2015), the battlefield had been rendered a *corpography*, requiring a different way of knowing, ordering and navigating the spaces of military violence.

It was also as many have pointed out 'a palimpsest of overlapping, multi-vocal landscapes' (Saunders 2003: 37). As the Western Front battlefield became, over time, a malign industrialised space where visibility was often a 'trap', so the military sketch was the spring in that mechanism. Concealment was the only antidote to the omnidirectional gaze of the trained eye.

Jay Appleton, developing Konrad Lorenz's thesis on the atavistic landscape, proposed a habitat theory that categorises any landscape into hierarchies of 'prospect, refuge and hazard' (Appleton 1975). The panoramic viewpoint is the paradigm of Appleton's system; military drawing systematised the graphic language so that trees became datum points, and the fixed features of the land became the immutable coordinates of a functional terrain, a strategic field. Or, as Henry Reed phrases it in this poetic fragment 'Judging Distances' from *Lessons of the War*, it is a domain where the temporal overlaps with the spatial:

> Not only how far away; but the way that you say it is very important./Perhaps you may never get/The knack of judging a distance, but at least you know/ How to report on a landscape: the central sector/The right of the arc and that, which we had last Tuesday/And at least you know/That maps are of time, not place, so far as the army/Happens to be concerned – the reason being/Is one which need not delay us. Again, you know/There are three kinds of tree, three only, the fir and the poplar, And those which have bushy tops to; and lastly/ That things only seem to be things.
>
> (Reed 1943: 155)

As a visual essay in spatial interpretation, the drawn artillery panorama has clear areas of jurisdiction. The foreground is considered irrelevant. To the gunner, the nearby is already controlled. The middle distance and the horizon are the essential and desired focal points. These, to borrow Appleton's phrase, are the prospect-rich domains and the most coveted. Furthermore, panorama drawings are predicated on trajectories and barrage lines. The horizon is the ultimate goal because it holds the promise of further territory for martial exploitation. During the First World War, the horizon took on special value when seen from the noisome mess of the front-line trench. Secreted in their observation posts, gunners described the green and unspoilt distance as 'The Promised Land' – perfect, but forever locked in an unattainable future.

These concerns, as W.J.T. Mitchell (1994) has observed, are the essential discourses of imperialism. Empires, according to him, move outward in space 'as a way of moving outward in time, the "prospect" that opens up is not just a spatial scene, but a projected future of "development" and exploitation' (Mitchell 1994: 16–17). The promise of control must permeate every level of military drawing. In contemporary drawing manuals, the single pencil line must be given the authority of military language:

> A line should be as sharp and precise as a word of command. A wavering line which dies away carries no conviction or information because it is the product of a wavering mind. Every line should be put in to express something. Start sharply and finish sharply. Press on the paper.
>
> (Newton 1915: 27)

Similarly, by ridding the page of any ambiguity or doubt, the panorama drawing aims to pre-ordain the future. This is mirrored in the written word of the military voice, which uses the active and instructive tense of command. It results in a language where the passive or conditional tense simply does not function; instead, 'Brigade *will* commence at . . . , Objectives *shall* be taken by . . . , reinforcements *will* be moved to . . . etc.'. (Keegan 1976: 266). Maps and charts drawn up before offensives bear a similar code; barrage lines are clearly marked in minutes of advance; in Northern France in June 1944, the objectives beyond the Normandy beachhead were marked out in time – D-Day plus one, plus two, etc. – as well as in space (Keegan 1976).

Instruction manuals in military sketching equate clarity of line with clarity of purpose. Ambiguity and doubt are (quite literally) ruled out. The margins of failure (like the estimated casualty rate) are clearly prescribed and then codified. Blank areas of the paper are not intended to be read as negative (or open) space as an artist might have understood it, but the area set aside for instructive wording. The panorama, though, could only make sense in a war where both sides were predominantly static, where a battlescape was shared but where the zones of control were clearly demarcated. The view from the opposing emplacements might be radically different, but the contested ground was rationalised and systematised using a shared vocabulary of grid and line. In his analysis of the 'tourism' of war, Jean Louis Deotte (in Diller and Scofidio 1994) has argued that the beachline of Normandy in 1944 constituted a common world, a shared objectivity for both defender (cooped in a concrete pillbox) and attacker (exposed in a metal landing craft). Both sets of adversaries experienced a 'reversibility of the points of view' because 'enemies share in common the same definition of space, the same geometric plane . . . they belong to the same world of techno-scientific confrontation, the substratum of which, here, is sight' (Diller and Scofidio 1994: 121).

Part five: coda

As visual artefacts, the military sketches and rolled panoramas are appreciably dull; as utilitarian objects they belong to the material culture of war. However, very few have survived. Unlike the longevity visited upon much of the art produced by official war artists, the military sketch or artillery panorama had no cultural status. They had only short-term usage, and they were often produced by talented amateurs or professional artists, architects, and surveyors many of whom felt they had to 'dumb down' to create what was needed. Many held the practice in low esteem. In fact, little had changed in this regard since the origins of the military academies (Army and Royal Navy) in England in the mid-eighteenth century (Martins 1999). In 1802, John Constable had roundly rejected the offer as a drawing master in one military academy fearing it would have been a 'death blow to all my prospects of perfection in the Art I love' (Constable, in Hardie 1966: 222). Other painters dismissed the practice as little more than 'mappy', decrying its 'tame delineation' of a landscape (Gough 1995: 63–64). Officially sponsored British war artists such as Nash, Nevinson and many others were not required to subvert their creative practice to such utilitarian work. Instead, their artistic practice was tested across a broader range of subjects and to deeper and more insightful ends. Their creative output set new benchmarks in the way that modern warfare could be presented, visualised, and understood. Their war work brought fresh (but lasting) insights into the onslaught on nature and on the radical shifts in the way space, time and motion could be re-presented and 're-membered'. Their creative renderings of war on the Western Front have now become an essential ingredient in the staple diet of its iconography.

And what of the military sketch? Has its use been made redundant with the advent of GPS, heat-seeking visual detection tools, enhanced aerial surveillance and so forth? One might assume so. And yet in recent wars, the military sketch has still been relied upon to provide essential coordinates of a battlefront. In the memoir-novel, *Jarhead*, of the first Gulf War, the sniper and his 'spotter' make small thumbnail sketches to help identify, isolate and pinpoint enemy targets (see Swofford 2003). In the Australian War Memorial Museum in Canberra, there are a number of preserved and framed range cards drawn in an amateur hand by an Australian artilleryman. They depict, with a practiced use of line and a reasonable attempt at indicative shading, the mountainous slopes held by the enemy in Uruzgan province, Afghanistan. 'Compiled' in 2008, these sketches were drawn to assist troop movements, to offer topographic information about potential enemy locations, direct-fire targets, and identify notable landmarks. The overlay of red pen indicates the extreme left and right hand of the firing arc and in action would help an infantry unit co-ordinate its defence. They may be rather artless but – as their artistic predecessors over the centuries would agree – they play an essential and seemingly timeless function in neutralizing dangerous terrain ahead and around them.

Bibliography

Abbot, C.C. and Bertram, A. (1955) *Poet and Painter: Being the Correspondence between Gordon Bottomley and Paul Nash, 1910–1946*, Oxford: Oxford University Press.

Alfrey, N. and Daniels, S. (1990) *Mapping the Landscape: Essays on Art and Cartography*, Nottingham: Nottingham Castle Museum.

Appleton, J. (1975) *The Experience of Landscape*, Wiley: London.

Bennett, A. (1918) *Void of War*, London: Leicester Galleries.

Bertram, A. (1955) *Paul Nash, the Portrait of an Artist*, London: Faber and Faber.

Bickel, L. (1980) *In Search of Frank Hurley*, Sydney, Australia: Macmillan.

Brown, M. (1978) *Tommy Goes to War*, London: J.M. Dent.

Carmichael, J. (1989) *First World War Photographers*, London: Routledge.

Cave, N. (1994) *Beaumont Hamel*, London: Pen and Sword Books.

Chasseaud, P. (2013) *Mapping the First World War: The Great War Through Maps from 1914 to 1918*, London: Collins.

Cloke, P. and Jones, O. (2004) 'Turning in the graveyard: Trees and the hybrid geographies of dwelling, monitoring and resistance in a Bristol Cemetery', *Cultural Geographies*, 11(3): 313–341.

Clout, H. (1996) 'After the Ruins': Restoring the Countryside of Northern France after the Great War, Exeter, UK: Exeter University Press.

Cork, R. (1994) *A Bitter Truth: Avant-Garde Art and the Great War*, New Haven: Yale University Press.

Diller, E. and Scofidio, R. (1994) *Tourism of War*, Basse-Normandie, FRAC Basse Normandie/University of Princeton Press.

Eates, M. (1973) *The Master of the Image 1889–1946*, London: John Murray.

Farrer, R. (1918) *The Void of War: Letters from Three Fronts*, London: Constable.

Foucault, M. (1975) *Discipline and Punish: The Birth of the Prison*, Paris: Editions Gallimard, English translation (1977), London: Allen Lane.

Gough, P. (1995) 'Tales from the Bushy-topped tree: A brief survey of military sketching', *Imperial War Museum Review*, 10: 62–74.

Gough, P. (2004) 'Sites in the imagination: The Beaumont Hamel Newfoundland Memorial on the Somme', *Cultural Geographies*, 11(3): 235–258.

Gough, P. (2009) 'Calculating the future', in N. Saunders and P. Cornish (eds.), *Contested Objects: Material Memories of the Great War*, Routledge: London, 237–251.

Gough, P. (2010) 'A Terrible Beauty': British Artists and the First World War, Bristol, UK: Sansom and Company.

Greenacre, F. (1988) *Francis Danby: 1793–1861*, London: Tate Gallery.

Gregory, D. (2015) 'Gabriel's map: Cartography and corpography in Modern War', in P. Meusburger and D. Gregory (eds.), *Geographies of Knowledge and Power*, Springer: New York, 89–121.

Hague, R. (ed.) (1980) *Dai Greatcoat*, London: Faber.

Hardie, M. (1966) *Watercolour Painting in Britain, Vol. 1: The Eighteenth Century*, London: Batsford.

Heffernan, M. (1995) 'For ever England: The Western Front and the politics of remembrance in Britain', *Ecumene*, 2(3): 293–323.

Hill, A. (1930) 'Artist at War', *The Graphic Newspaper*, 15 November, London: 31–32.

Jones, D. (1937) *In Parenthesis*, London: Faber.

Keegan, J. (1976) *The Face of Battle*, London: Penguin.

Kern, S. (1983) *On the Culture of Time and Space 1880–1918*, London: Weidenfeld and Nicholson.

Malvern, S. (2001) 'War tourisms: "Englishness", art and the First World War', *Oxford Art Journal*, 24(1): 47–66.

Martins, L. (1999) 'Navigating in Tropical Waters', in D. Cosgrove (ed.), *Mappings*, London: Reaktion, 148–168.

Mitchell, W.J.T. (1994) *Landscape and Power*, Chicago: University of Chicago Press.

Nash, N. (1948) Outline: An Autobiography and Other Writings, London: Faber and Faber.

Newton, W.G. (1915) *Military Landscape and Target Indication*, London: Hugh Rees.

Redgrave, R. (1886) *A Century of Painters*, London: Graves and Armstrong.

Reed, H. (1943) 'Judging distances', *New Statesman and Nation*, 25(628): 155.

Roberts, W. (1974) 4.5 Howitzer Gunner RFA: The War to end all Wars, London: Canada Press.

Saint-Amour, P. (2003) 'Modernist reconnaissance', *Modernism/Modernity*, 10(2): 349–380.

Saunders, N. (2003) *Trench Art: Materialities and Memories of War*, Oxford: Berg.

Saunders, N. (2014) *The Poppy: A History of Conflict, Loss, Remembrance & Redemption*, London: Oneworld.

Sebald, W.G. (1995) *The Rings of Saturn*, London: Harvill.

Stuart Dolden, A. (1980) *Cannon Fodder*, Blandford, UK: Blandford Press.

Swofford, A. (2003) *Jarhead: A Marine's Chronicle of the Gulf War and other Battles*, Scribner: New York.

Talbot Kelly, R. (1980) *A Subaltern's Odyssey: A Memoir of the Great War, 1915–1917*, London: William Kimber.

Weir, B. (2007) '"Degrees in nothingness": Battlefield topography in the First World War', *Critical Quarterly*, 49(4): 40–55.

Wylie, J. (2007) 'The spectral geographies of W.G. Sebald', *Cultural Geographies*, 14(2): 171–188.

3 Remembering the anti-war movement

Contesting the war and fighting the class struggle on Clydeside

Paul Griffin

At Christmas 1915, the spirit of rebellion both against the continuance of the War and against working and living conditions was so strong that Mr Lloyd George accompanied by Mr Arthur Henderson paid a special visit to Glasgow to try to pacify workers (James Maxton Diaries, 1915–16).[1]

Introduction

The anti-war movement on Clydeside had a presence in the city from the declaration of war with 5,000 people attending a peace demonstration held at Glasgow Green in August 1914.[2] Such contestation of the First World War must form a central part of its remembrance and this chapter will primarily draw upon examples from a Scottish context to engage with anti-war sentiment that took place throughout the war. The chapter asserts that activists and organisations within Glasgow, and the broader Clydeside region, were central in articulating an anti-war message. The opening quotation from ILP member and future leader James Maxton is indicative of this sentiment and reveals connections between class antagonism and the anti-war movement.

Rather than a more general focus on conscientious objection, this chapter pays attention to the political opposition to the war, given that this aspect was particularly notable within the Clydeside context.[3] In this regard, the resistance during this period formed a central part of the period now remembered as Red Clydeside[4] whereby Glasgow, and surrounding towns and villages along the River Clyde, became largely defined by socialist values, political resistances and the growth of the trade union movement (see Damer 1990; Brotherstone 1992; Griffin 2015a). ILP member and anti-war activist Fenner Brockway commented on the distinctive nature of Clydeside's opposition to the war, suggesting that its relative success (in comparison to the London-based leadership) was due to its dual focus on the class struggle and peace movement. He argued that 'they were speaking a different language . . . [w]e concentrated on peace. They concentrated on the class struggle' (Brockway 1942: 53). This chapter argues that the relationship between the anti-war movement and class politics manifested itself in multiple ways. This unique combination of class and political antagonisms allows the chapter to narrow its focus to the particularly hostile response of Glasgow to the war, and to assert the broader significance of remembering those who committed themselves to contesting the war. By doing so, the chapter seeks to broaden understandings of 'peace' movements by indicating the diverse motives for contesting the First World War.

To provide an overview of the diverse activities, which forged the 'spirit of rebellion' described by Maxton, the anti-war movement is considered through the lives of three politically active individuals. The analysis cross-references their anti-war biographies to characterise different strands of the anti-war movement within Glasgow. The experiences of Guy Aldred, Helen Crawfurd and James Maxton were reflective of the diverse support for the movement. Their activism during the war is considered here as being representative of the plurality of political positions within the anti-war cause and is illustrative of the strong connections between the anti-war movement and broader class struggle. Their lives are representative of the multiple currents, specifically the anarchist tradition (Aldred), suffrage and feminist movements (Crawfurd), and parliamentary left activism (Maxton), which combined in opposition against the war. In this regard, these activists, as part of the broader movements discussed, formed a strong part of the longer trajectories of labour and political organising associated with Red Clydeside. Most importantly, for the remembrance of the anti-war movement, this combination of positions is indicative of the strength, depth and diversity of the communities that formed the broader movement.

Throughout the analysis that follows, the chapter proposes that a more general sense of dissatisfaction with the consequences of war must be read alongside the histories of explicitly anti-war movements. This more comprehensive sense of discontent reflects the dynamism and discontinuities of labour histories. Such collaborations and tensions are central to how labour geographies and histories are written (see also Featherstone and Griffin 2016). In particular, their foregrounding allows a critical review of the different strands of political struggles that connect different strands of the anti-war movement. This facilitates the inclusion of a much broader understanding of what constitutes labour and social movements, through the inclusion of aspects such as anarchism and gendered activism, to engage with connections and relations between struggles rather than more singular histories. Further, this approach stresses connections beyond place-based disputes to foreground the making of translocal connections within labour and social movements (see also Featherstone 2012).

The chapter is structured around three key geographical aspects of the anti-war movement on Clydeside. It begins by engaging with the variety of protest activity following the outbreak of the war. This analysis centres upon the multiple articulations of anti-war sentiment as evident through the lives of the three political activists. A central geographical contribution here is to argue that key sites and documents emerged from these practices that provide a spatial politics to the movements, which still resonate in the present day. The second section pays closer attention to the connections between particular class antagonisms and the anti-war movement. It does so by considering overlaps between the rent strikes and anti-war movement whilst also considering links with work-place disputes. This analysis develops the previous section by again indicating the diverse support for opposition to the war and how the activists introduced produced multiple related demands that stretched beyond pacifism. The final section briefly considers the broader links between these individuals and anti-war movements elsewhere to illustrate the translocal and international spaces of solidarity that developed during this period. The chapter concludes by drawing these themes together and briefly reflecting on more direct links to contemporary anti-war demonstrations.

The anti-war movement on Clydeside

The anti-war movement was active on Clydeside from the outbreak of war. It was aligned with broader social, political and industrial unrest that defined Red Clydeside. Meetings, marches and demonstrations of varying sizes took place throughout the war and the *Forward* newspaper, published on behalf of the ILP, became a key document for circulating meeting information and critiquing the war.[5] It also reflects the geographies of the movement and by virtue of being a document which would have circulated amongst people during the war provides socialist interpretations of war news. Damer (1990: 119) has described the newspaper as 'open to all groupings of the left, including the considerable anarchist presence in Glasgow and the Scottish nationalists'.

Anti-war demonstrations and general restlessness towards the consequences upon living standards were reflected during marches shortly after the declaration of war. The *Forward* described the march in August 1914:

> The gathering was Cosmopolitan in character and included doctors and dock labourers and rebels of every possible brand from mild peace advocates to the wildest of revolutionaries.
>
> (*Forward*, 15/8/14)

The diverse political reasons for opposition to the war were central to Clydeside's solidarity in contesting it. At the end of 1915, there were anti-conscription rallies in Glasgow, ILP conferences to discuss their position on the war and broader practices of community organising throughout the city. A sense of international solidarity and a workers' critique of capitalism was central to these anti-war movements and this was further evident during the demonstration highlighted earlier, particularly the 'Peace Society platform':

> Councillor Taylor read resolutions passed by huge gatherings of workers in Austria, Germany and France against war.
>
> The resolution passed:

> That this meeting deplores the outbreak of War and declares it to be the outcome of Capitalism allied with Militarism, which has been consistently opposed by the organised workers and pacifists in all the countries concerned and this meeting sends fraternal greetings to the working classes of and

pacifists of Germany, Austria, Italy, Russia, France, Belgium, Servia, and all other countries.

(Forward, 15/8/14)

This spirit of internationalism was prominent within Clydeside's opposition to the war and is further reflected in a later section of this chapter. More broadly, the diversity of political positions within this anti-war movement is perhaps best represented through the activities of Aldred, Crawfurd and Maxton. Their activism during this period is indicative of the diverse reasoning for discontent within Clydeside during the war but also illustrative of the different solidarities that emerged from this period. Some examples of their political activism during this period are raised next to illustrate the diversity of support for the anti-war movement on Clydeside, which fed into the protests, such as those raised earlier, during the First World War.

Helen Crawfurd channelled her anti-war efforts through the Women's Peace Crusade (WPC), as secretary, alongside other Glasgow based women (including Mary Barbour, Agnes Dollan and Ethel Kaye). These women had a significant presence within the city, holding regular open-air meetings around Clydeside and producing leaflets, pamphlets and badges that were distributed throughout Scotland. Such activities were representative of the strong female position within the working-class presence of Clydeside (Breitenbach and Gordon 1992). Reflecting on her involvement with this movement in her memoirs, Crawfurd stressed the importance of the street corner as a political meeting place during this period and these more unofficial meetings, when compared to the 'official' trade union forms of organising, were central to WPC organising.

An example of this organising occurred in July 1917, when the WPC organised a demonstration of 5,000 people, which Couzin (2006: n.p.) has described through engagements with newspaper reports as a 'mass demonstration in Glasgow; from two sides of the city processions wound their way through the city accompanied by bands and banners'. Such processional cultures were central to the formation of group identities and collectives as Harvey et al. (2007) have shown with Methodist processions in the late nineteenth century. On their arrival to Glasgow Green, they then 'merged into one massive colourful demonstration of some 14,000 people'. These demonstrations engaged with urban landscapes and, as Navickas (2009: 93) has highlighted in relation to popular protests in Yorkshire and Lancashire, such places 'contributed to this extraordinary atmosphere as both venue and as symbol'. Glasgow Green, for example, became a key site for the working-class presence as a place of many demonstrations and as a site representative of freedom of speech its accessibility eventually became a key campaign fought for by Guy Aldred.

During 1914, Crawfurd wrote 'Our Suffrage Column' for *Forward* in which she highlighted the need for a working-class solidarity to confront the capitalist war. In the extract that follows, she connects the women's suffrage movement, which she was strongly committed towards, with the challenging of the capitalist notion of international war:

We women, who are looking for the dawning of that day when wars shall cease, believe that with liberty for women that day draws nearer. Today one thinks of the forcible feeding that the educational authorities perform upon the youth of this country, filling their minds with what the Capitalists call "patriotism". War is glorified and international hatreds developed and encouraged. Submission and discipline encouraged, even if that submission means submission to injustice.

(Forward, 22/2/1914)

This gendering of the challenge towards patriotism and the war was a crucial part of the working-class presence during this period and contested more masculine articulations of the broader labour movement. Crawfurd's anti-war stance eventually ended her previous political loyalty to the Women's Social and Political Union, who she described as supporting the war, as she joined the anti-war ILP alongside Maxton (Crawfurd n.d.).[6] Such porous political groupings and fluid political identities were a common feature of this period.

Due to the political pressure emanating from Glasgow in 1916, the *Forward* newspaper, which had regularly posted 'Socialist War Points' and anti-war propaganda, was suppressed from publishing by the British government. The paper was suspended for a month and only resumed publication after agreeing to publish nothing to cause 'disaffection with the 168 Munitions of War Act or with the policy of the dilution of labour'.[7] The suppression of such newspapers was a major setback for the left on Clydeside and was also indicative of the level of threat perceived by the British political establishment, illustrating the importance of Glasgow in articulating the anti-war message.

James Maxton was connected to the *Forward* through his membership with the ILP.[8] His anti-war activity was prominent from the outbreak of war and was representative of parliamentary opposition and further overlaps between different strands of the labour movement. ILP opposition was largely centred upon a critique of the war as profiteering and Maxton would articulate this from the platform at demonstrations. This opposition was central to the growth of the ILP, with Kenefick (2007: 133) stating that by the Armistice of 1918, the number of Scottish branches had more than doubled and membership was at three times its pre-war level. The *Forward* newspaper reflected a political-economic critique of the war, as it continually stressed the economies of war and the reality of 'rich capitalists' benefitting from profiteering at the expense of the working class in its weekly update of the war. At a conference of Scottish Miners in 1915, Maxton made this point himself stressing that the war had provided 'the speculators a unique opportunity of plundering the workers' (cited in Knox 1987: 19). Such opposition brought Maxton into contact with the No Conscription Fellowship (NCF) and his challenging of the Military Service Act, which led to his eventual arrest. The NCF was an organisation founded primarily by Liberals but had a diverse supporting group to contest compulsory military call-ups.

The importance of an explicit political opposition to war, rather than moral or religious groundings, was central to Clydeside's opposition to the war. In

Glasgow, socialist and anti-war campaigner John Maclean stressed that it was important to 'get jail politically', rather than as a conscientious objector on pacifist grounds, and this broader political message appeared important to the three activists considered here.[9] Maxton and Guy Aldred were both imprisoned due to their political opposition. Maxton was charged with sedition following public speaking appearances whilst Aldred was charged with failing to report for military service. Aldred was held in prisons and military camps before being sent to a labour camp in the village of Dyce, near Aberdeen (see Caldwell 1988). As has been well documented, prison conditions and labour camps were particularly harsh for anti-war agitators or conscientious objectors and could lead to mental health issues (see Brown 1986). These experiences and forms of opposition were not necessarily unique to Glasgow, with political resistance prominent throughout the United Kingdom, but were arguably more diverse due the incorporation of different strands of political resistance during this period.[10]

One such strand of opposition that can be overlooked in remembering the anti-war movement is the broader critique of war that emerged from anarchist and socialist libertarian positionalities. Within the Clydeside context these are best represented through the life and writings of Guy Aldred. Aldred was based in London during the war but he travelled regularly to Scotland to share platforms with several prominent Glasgow activists (he would eventually move to Glasgow permanently following 1919). His publications and public speaking were central in articulating alternative visions of what the anti-war movement meant. The content of his public appearances within Glasgow are difficult to trace but can be positioned within his broader political activism. His opposition to militarism was forged on an anti-parliamentarian spirit, with Aldred retrospectively describing his motivation as adopting the following:

- Anti-patriotism: complete refusal to assist militarism or navalism of any kind.
- No peace by negotiation. Incessant war against all Capitalist Governments. Refusal to bear arms on class-conscious grounds.

(The Commune, March 1924)

These comments, in one of Aldred's many publications, reflected his position on war and are specifically positioned against pacifism which he described as 'privileged conscientious objection and peace by negotiation'. Such views were likely articulated during his many public speaking appearances within Scotland. In a biography of Aldred, J. T. Caldwell (1988: 143) documents how he had 'total disregard for labour camp rules' in Dyce and still managed to attend and speak at meetings in Glasgow. He was a regular speaker at many meetings during 1916 and continued to articulate an alternative vision of the anti-war movement as detailed next. This engagement with Glasgow continued throughout the war and Aldred was recorded as a speaker for the May Day demonstration in 1919 (*Forward*, 26/4/1919).

This anarchist strand of political resistance, whilst clearly critical of many elements of the overall anti-war movement, was a key competent of the discontent on Clydeside. This political positionality must be included within a historical account of the political motivations for opposing the war. Whilst Aldred was not a constant presence on Clydeside, anarchist stalls were consistently active at demonstrations. An example of an activist with connections to Aldred was Glasgow based anarchist John Smith, who was arrested in a similar manner and for the same reasoning as Maxton in 1916. His sentencing was six months longer though and it was argued that this was due to his association with Aldred with the court linking him to a 'well-known London anarchist' (Aldred is meant here). He was also recorded as carrying eight copies of a paper described as being of 'extreme character' that contained a statement from the International Stop the War Committee which advocated the ending of the war through revolution, and the spreading of communism (*Western Daily Press*, 12/5/1916 and *Derby Daily Telegraph*, 11/5/1916 cited in Heath n.d.). Such connections are indicative of the centrality of Aldred's contribution to the Clydeside movement, the importance of activist connections beyond Glasgow and the prominence of an anarchist thread within the anti-war collective.

These connections and solidarity between different strands of the left was also commented upon by Maxton who suggested that political differences could be overcome during particular disputes:

> I never have been in the same Party as either Maclean or Aldred, although frequently appearing on platforms with them when particular fights on special issues made such alliances the obviously right course to take.
>
> When a common platform was agreed upon to secure unity of working class outlook on a particular issue, one could go on to the platform with full confidence that they would deal with that issue, and struggle to get a united workers' mind on the subject.
>
> (Maxton cited in Aldred 1944: 107)

This combination of activists and organisations illustrates the depth of the anti-war movement and indicates the solidarity forged upon the issue. Despite their political differences, all three political activists contributed towards a broader anti-war movement on Clydeside. Maxton and Crawfurd were also regularly recorded as speakers at the same meetings such as the 'no conscription' demonstrations in 1915. These combinations of different activists, such as the solidarities between the three activists considered here, were a key component of the anti-war movement. The chapter argues that this diversity and solidarity must be acknowledged in the remembrance of such movements.

This overall strength of this movement was perhaps reflected best in 1918 when protests occurred during May Day demonstrations. Over 100,000 were present during this demonstration, which included a platform appearance from James Maxton. The *Forward* (11/5/1918) newspaper described these scenes as Labour's Greatest May Day with 'Lively Scenes: Capitalists Enraged' and:

> twenty odd platforms on Glasgow Green, a galaxy of orators, tons of Socialist literature sold, and the capitalists driven nearly crazy with rage at the spectacle of it all.

This was clearly a momentous event for the labour movement and the anti-war sentiment throughout the event was strong. One Co-operative lorry held a banner, which read 'MAN'S INHUMANITY TO MAN MAKES COUNTLESS THOU-SANDS MEATLESS' and many other platforms held anti-war speeches. These scenes were not completely united though and again demonstrators received a hostile reception from some onlookers. Shouts of 'This is no time for a holiday!', 'Go back to your work you stalkers' and 'Away and join the army' were recorded in the paper cited earlier and further illustrated the difficulty in opposing war. Despite these hostilities, the demonstration remained strong; with a 'great crowd' recorded as going to Duke Street Prison to support the imprisoned John Maclean in the hope that he would hear their chants of his name.

Such evidence of anti-war activities from the outset of war through to the May Day demonstrations in 1918 indicates a longevity to a movement that contested the war throughout its duration. All three activists mentioned were central to articulating an alternative vision of the war to the propaganda emerging from government. Thus, the anti-war movement had a clear presence on Clydeside. This presence was notable through publications, such as the *Forward* and *The Spur*, more transitory strategies such as the badges and meetings of the WPC, and the larger-scale rallies at places such as Glasgow Green as detailed earlier. These strategies and places of protest were central to the geography of the anti-war movement and still resonate as sites and methods of organised resistance today. They also reflect the difficulty in articulating an anti-war message, through both suppression from the state and antagonism from those supporting the war effort.

Connecting struggles: anti-war sentiment and class antagonism

As noted earlier, a key aspect of the anti-war movement on Clydeside was the connections made between class and peace movements. In this regard, Crawfurd's involvement with the anti-war movement was also indicative of broader connec-tions between anti-war sentiment and struggles regarding social reproduction and rising rents (see also Gray 2015). This issue became highly contested on Clydeside during 1915 and led to one of the most notable and successful campaigns of the Red Clydeside period. In their analysis of the role of anarchism and syndicalism in contesting the war internationally, Schmidt and van der Walt (2009: 211) claim that '(a)narchist and syndicalist opposition did not derive so much from pacifism – an opposition to violence in any form – but from a class analysis'.

During November 1915, households across Glasgow refused to pay rents to landlords and prevented tenant evictions through organising within tenement communities. Many WPC women were key contributors to these rent strikes during 1915–16 (for example Mary Barbour and Agnes Dollan). This connection provides further evidence for Brockway's initial point about Glasgow's combined efforts to contest war and continue existing forms of class struggle. Maxton was also linked to the rent strikes (see Brown 1986) through his involvement within the ILP, who negotiated much of the struggle politically, but as Melling

(1983) indicates, the rent strikes were primarily a women led movement. The rent strikes are often referred to as the most successful campaign to emerge from the Red Clydeside period. The campaigners, primarily women within Glasgow Women's Housing Association, including Crawfurd, forced the government into imposing a rent restriction act that returned rent costs returning rent to pre-war rates. Crawfurd's own role within these events is also acknowledged by Melling (1983), and was perhaps most notable in the scenes during an eviction hearing at a Glasgow court:

> Mrs Crawfurd said that this fight was essentially a women's fight. All who were taking part in the demonstration were showing their solidarity. They were not asking for mercy, not for charity; they were asking for justice. When the Government brought in the moratorium at the beginning of the war they could have made it illegal for factors to increase rents and for bondholders to raise the interest on their bonds.
>
> (*Glasgow Herald*, 15/11/1915, p. 11)

The opening line of this quote speaks to Smyth's (1992) assertion that the gendered aspect of the rent struggle should not be subsumed with a broader labour history. The strike was led by women and reflected issues largely experienced by women. However, the gathering of campaigners outside the court hearing was illustrative of a broader solidarity amongst the working class in Glasgow on the issue of rent that connected with other wartime struggles. During this hearing, members of engineering and shipbuilding unions also attended as part of sympathetic industrial action (see Royle 2006) in support of the rent campaign. These relationships between industrial workers and those involved with the rent strikes were largely maintained during the war and overlapped with the industrial demands that are considered further next. Their campaigns, publications and actions all contributed towards resistances during the war and illustrated a multiplicity and diversity within related demands.

Maxton's diaries also highlight many of the labour and trade union disputes during the war. They document the famous response of the shop stewards to the visit of Lloyd George (Minister of Munitions at the time) on Christmas Day in 1915. The scenes at the meeting on Christmas Day are well documented – described by the *Forward* (1/1/1916) newspaper as 'wild scenes' where 'Mr Lloyd George was received with loud and continued booing and hissing' – and illustrated the growing frustration towards the effects of the war on Clydeside (see also Craig 2011). The meeting was arranged to gain support for an agreement on changes to wages and working conditions. Unskilled labourers, often women, had been brought in to munitions factories and were viewed by the established engineering unions (in particular the Clyde Workers Committee) as a threat to the position of skilled labour. Labour disputes were perhaps more prominent in Glasgow than anywhere else in the United Kingdom during the war, with workers at the Fairfield yard contesting the imposition of leaving certificates, union organisers attempting to organise women introduced at Beardsmore and labour disputes at workplaces such as Weir's in Cathcart (see also Foster 1990). These industrial responses to the war were a central part of Clydeside radicalism during this period.

With these industrial disputes in mind, it is important to stress that the class movement on Clydeside was not always entirely united against the war, as the disputes of unions such as those of the Clyde Workers' Committee were often over industrial conditions rather than the war itself. This tension was openly exposed by Guy Aldred and socialist-libertarians who made a distinction between disputes over wartime conditions and anti-war arguments. Aldred specifically targeted munitions workers arguing that for 'every bullet you make to kill a German soldier is aimed at the heart of your conscript son' (Aldred cited in Caldwell 1988: 117). His comments had undoubted relevance, with ley labour organiser David Kirkwood pledging, despite his central role organising workers to contest dilution, that he 'must be devoted to my country' (cited in Royle 2006: 132). This pledge to continue production of munitions but with improved labour conditions was thus not a critique of war itself but rather articulated an opposition to wartime conditions. Thus, there were often clear limitations to trade union based critique of the war, which whilst contributing towards an overall spirit of rebellion and discontent, remained essentially labourist and primarily concerned with dilution struggles. This defence of labour rights and protectionism regarding skilled jobs, due to the introduction of women workers and skilled American engineers who were receiving better pay, were central to the broader sense of general discontent during the war. These complaints were also framed within a broader critique of a growing cost in living during the war, which, although not explicitly anti-war, were forged in the face of being described as anti-patriotic in similar terms to the anti-war protestors during the May Day demonstrations. Here, it is argued that this broader dissatisfaction with the war was a key part of the sense of opposition to the war and therefore central to the anti-war movement on Clydeside.

Spatial politics: translocal connections and the anti-war movement

The importance of political activists such as Crawfurd, Maxton and Aldred in maintaining and developing Clydeside's anti-war movements has been indicated earlier. Such political antagonisms should be considered in relation to the position of the broader European socialist left during the First World War.[11] Eley (2002: 125) has highlighted how the political left became particularly weak during this period as 'recognizing the International's powerlessness, socialists rapidly moved into actively supporting the war'. This often resulted in 'national defencism' and a radicalisation of labour often based on terms that supported the war effort. However, as is acknowledged by Eley, Glasgow provided a significant contrast to this position with a radical political economy perspective firmly opposed to the war and explicitly anti-imperialist. This left political position, therefore, maintained a critique of militarism and its consequences throughout the war, which was grounded within international notions of social justice. Partial reasoning for this was due to changes within the workplaces and living conditions of Glasgow but as is argued here, the opposition was also as part of a broader movement that linked war resistance to a criticism of 'capitalism and the state' (Schmidt and van der Walt (2009: 215).

The anti-war movement had a strong sentiment of internationalism as its basis with Maxton, Crawfurd and Aldred holding strong views on the importance of internationalism to the socialist movement (however defined). Crawfurd described her God within her memoirs as an international one. Aldred continually emphasised the importance of internationalism in his papers and stressed the importance of his 'postal mission' (Aldred 1963) to connect with comrades. Whilst Maxton, in a prison letter to his mother, described how he had 'perhaps erred from the patriotic point of view, but not from the wide-world humanitarian view'.[12] Such expression of solidarity with comrades was demonstrative of the international imaginaries of political activists during this period. It was prominent within the anti-war movement coverage within the *Forward* and is reflected in the lives of the three political lives considered here. This international dimension is crucial for rethinking labour histories as being inclusive of imaginaries beyond immediate circumstance (see Featherstone 2012).

Maxton received many letters during a significant illness in 1926 and then, after his death in 1946, which reflected his connections across Europe and beyond. The content in these letters varies but his opposition to the war is a clear theme to many of them. One letter from Jamie and Hein Van Wijk, Haarlem, praised Maxton's internationalism, describing him as representing 'the best traditions of international socialism', and recognising him specifically for his passionate voice against the First World War.[13] Prior to his own imprisonment under the 1914 Defence of the Realm Act, Aldred wrote a short article entitled 'Meditation', published in *The Spur*. This article was translated by French anarchist Emile Armand and republished in his paper, "Par dela la Melee" in February 1917. The article was written the day before Aldred's court appearance and was later reproduced in further French anti-militarist papers and circulated in leaflet form by French anti-war campaigners. Such correspondences begin to reflect the contrasting influences within Clydeside and also the connections forged by the activists considered in this chapter. Thus, the political actions and correspondence made Glasgow and its associated networks a key site within the anti-war movement. These connections begin to illustrate a combination of political activists and international communications within and beyond the working-class presence of Clydeside.

Aldred commented on the political significance of such correspondences in developing this position, when describing the exchange of pamphlets and writing:

> It possessed the charm of penetration into unknown territory. There was a touch of mystery about such activity. One could not see what would result from the mere putting of a pamphlet, duly stamped, into a post box, and thus sending it by unknown hands to an unknown person. It was like performing a miracle.
>
> (Aldred 1963: 429)

This account of circulating documents is particularly relevant to early twentieth century Clydeside and the anti-war movement in particular. The exchange of news and papers forged material connections between places and activists. Such translocal solidarities were particularly prominent on Clydeside and were consistently articulated by the three individuals. They are indicative of geographical networks of solidarity and begin to reveal the significant role played by Clydeside activists in distributing an alternative vision of the war.

Crawfurd's activism was similarly forged through translocal connections and a commitment to internationalism. Following the start of the war, Crawfurd became increasingly connected to events in Ireland and was influenced by the Irish radical figure James Connolly, with whom she was in contact with during her many visits to Belfast to speak on behalf of the ILP. In her memoirs, she reflected on Connolly's description of how 'Ireland will not have the conscription. Let the law say what it likes. We have no intentions of shedding our blood abroad for our masters' (Crawfurd n.d.: 117). As a member of the ILP, and on behalf of Tom Johnston (editor of *Forward*), she would convey messages on her travels to Belfast. During her time in Ireland, she also developed important connections with radical Irish women including Constance Markiewitz, Maud Gonne and Charlotte Despard. These connections and friendships influenced Crawfurd's activities within Glasgow and were reflected in her political activism and through publications such as *Forward*.

The brief examples raised earlier have illustrated the commitment of the activists considered to a broader to international socialism within their political opposition to the war. This commitment was thus not restricted to Clydeside and these connections suggest a broader anti-war movement that suggest that Clydeside's opposition was forged discretely. More broadly, the internationalism element of political radicalism was also maintained by each individual in the following years. In the 1920s, Maxton was a central figure within the League against imperialism and received praise from Pan-African journalist and political organiser George Padmore for his opposition of government restrictions placed on the *Negro Worker* publication (see Pennybacker 2009: 68). Helen Crawfurd visited Russia in 1921 and interviewed Lenin, before becoming secretary for Workers' International Relief. This was a Comintern organisation (O'Connor 2004: 7), which supported workers during strike activity by providing essential resources. Similarly, Guy Aldred continued to hold many international friendships and was particularly active during the Spanish Civil War (see Caldwell 1988). This internationalism was central to their opposition to the war and subsequent political lives. It also positioned these activists and radical groups on Clydeside as central to articulating alternative war imaginations.

Conclusion

Anti-war sentiment during the First World War illustrates clear intersections between radical figures and traditions on Clydeside. The combination of class and anti-war struggles provides the most explicit example of Brockway's (1942) claims regarding the strength of Clydeside's anti-war movement. The movement clearly combined class interests (or economic demands) with wider political demands and discussion. This combination of economic and political struggles was integral to radical Clydeside's spirit. It is also notable that this period of radicalism was one of the most united periods of organising with little infighting within the left on Clydeside (see also Griffin 2015b). As Brockway has claimed, the strength of the anti-war war movement on Clydeside was seemingly more prominent than elsewhere in Britain and its strength emerged through its combination of diverse political positions. These anti-war efforts, whilst primarily viewed through the lens of three political personalities, were also clearly reflective of a much wider community.

This chapter has argued for the recognition of difference within spaces of remembrance. The First World War prompted a diversity of responses within Glasgow, which mixed explicitly anti-war opposition and protests against wartime conditions. The chapter has considered this general discontent to challenge understanding of what constitutes histories such as the anti-war movement and Red Clydeside more specifically. The anti-war biographies of political activists such as Aldred, Crawfurd and Maxton are representative of the diverse political opposition to the war. Whilst engaging with this relational construction of a Red Clydeside narrative, it remains important not to over emphasise the novelty of the Glasgow experience, as anti-war opposition was evident elsewhere, including elsewhere in Scotland,[14] and the activists considered here were clearly well connected with similar struggles. Thus, the political lives considered can also be positioned within a longer trajectory of anti-war campaigning and a broader challenging of perceptions regarding war (see Mueller 1991).

In this regard, it is notable that Red Clydesiders are still referred to within contemporary struggles and that more recent anti-war movements have replicated strategies used a century ago. Places such as Glasgow Green and George Square remain key organising sites for protest movements and reflect a history of radicalism on Clydeside. This longer trajectory of radicalism was evident in 2003 during the anti-Iraq war march that gathered at Glasgow Green, the site of many of the protests referred to earlier, before marching to the Labour Party conference in Glasgow (see BBC 2003). These connections between past and present are critical to the geography of anti-war movements (see also Gillan et al. 2008) and shows the prominence of places and connections within their making. Such an approach to the solidarity emerging from the anti-war movement produces a more open account of protest geographies and politics as well as a more relational account of labour history that brings together diverse strands of resistance.

Notes

1 Glasgow Mitchell Library, TD 956/6/15.
2 Glasgow Green is a park in the east end of Glasgow, on the north bank of the River Clyde. Established in the fifteenth century, it is the oldest park in the city. During the

early twentieth century, it became a key organising space for protest movements and was subject to freedom of speech disputes. This direct action was co-ordinated by a combination of the Independent Labour Party (ILP), the British Socialist Party and the Glasgow Branch of the Peace Society.

3 Kenefick (2007) engages with conscientious objection more closely but also indicates how three-quarters of these were 'political objectors'.

4 A label attributed by a journalist according to McLean (1983).

5 This chapter uses this source extensively to provide a thorough and overarching account of activities during this period.

6 The war was divisive within the WSPU as the official leadership, primarily through Emmeline Pankhurst, supported the war efforts and 'gave short shrift to those on strike, slackers, conscientious objectors and pacifist whose unwillingness to contribute to the war effort was contrasted with the loyal war service of women' (Purvis 2007: 148). This resulted in many women, including Emmeline's daughter Sylvia, leaving the organisation to pursue anti-war efforts.

7 ILP Annual Report, 1916:14. London School of Economics, ILP 12/1/3.

8 Maxton would later go on to represent the ILP in parliament and chaired it twice between 1926–31 and 1934–39.

9 John Maclean was arguably the most notable of the Red Clydeside activists during this period. He was a central figure within the anti-war movement, described as a 'one man revolutionary party' (Damer 1990: 130) and his life has been document in biographies and detailed labour histories. His anti-war activism was central to Glasgow's response which culminated in his arrest and imprisonment (see Milton 1973; Royle 2006). Harry McShane Interview held at Gallacher Memorial Library, Glasgow Caledonian University, No date, Uncatalogued.

10 Cyril Pearce (2001) for example highlights political opposition as a significant part of his historical study primarily based in Huddersfield, see *Comrades in Conscience*.

11 German opposition to war was prominent in July 1914 with large demonstrations (30,000 were recorded in Berlin) taking place across the country but by August German and French parliamentary Socialists voted for war. Socialists in Belgium, Austria and Hungary took a similar stance by adopting 'national defencism'. Socialist opposition emerged from the Serb, Russian, Italian and Bulgarian Socialists (Eley 2002).

12 James Maxton Letters – 31/3/1916. Glasgow Mitchell Library, TD 956/4/2.

13 Another letter received from Mexico reflected Maxton's longer commitment to anti-war movements and detailed a previous meeting in Paris where the collective was attempting to save a party of Spanish revolutionists and praised 'the precious aid' Maxton gave to the cause. Similar letters were received from America, France and Germany. James Maxton Letters, Glasgow Mitchell Library, TD 956/4/2.

14 It is important to note that the *Forward* documented significant anti-war protests in Edinburgh in Dundee, such as those during May Day of 1918.

Bibliography

Aldred, G. (1944) *A Call to Manhood and other Studies in Socials Struggle, 26 Essays by Guy A. Aldred*, Glasgow University Special Collections, Broady Collection, I1.

Aldred, G. (1963) *No Traitor's Gait, The Autobiography of Guy. A. Aldred*, Glasgow University Special Collections, Broady Collection, I 22 – I40.

BBC (2003) *Organisers Hail Anti-War Protest*. [Online]. Available from: http://news.bbc.co.uk/1/hi/scotland/2765093.stm (accessed 18 December 2015).

Breitenbach, E. and Gordon, E. (1992) 'Introduction', in E. Breitenbach and E. Gordon (eds.), *Out of Bounds: Women in Scottish Society 1800–1945*, Edinburgh: Edinburgh University Press, 1–9.

Brockway, F. (1942) *Inside the Left*, Nottingham: Spokesman.

Brotherstone, T. (1992) 'Does Red Clydeside really matter any more?', in R. Duncan and A. McIvor (eds.), *Militant Workers: Labour and Class Conflict on the Clyde 1900–1950*, Edinburgh: John Donald, 52–80.

Brown, G. (1986) *Maxton*, Edinburgh: Mainstreaming Publishing.

Caldwell, J.T. (1988) *Come Dungeons Dark: The Life and Times of Guy Aldred, Glasgow Anarchist*, Ayrshire: Luath Press.

Couzin, J. (2006) *Radical Glasgow*. [Online]. Available from: www2.gcu.ac.uk/radical-glasgow/chapters/index.html (accessed 31 March 2015).

Craig, M. (2011) *When the Clyde Ran Red*, Edinburgh: Mainstream Publishing.

Crawfurd, H. (no date) *Unpublished Memoirs*. Original version held at Marx Memorial Library and Worker's School, HC/1/3. Uncatalogued copy consulted at the Research Collection at Glasgow Caledonian University.

Damer, S. (1990) *Glasgow: Going for a Song*, London: Lawrence and Wishart Limited.

Eley, G. (2002) *Forging Democracy: The History of the Left in Europe, 1850–2000*, Oxford: Oxford University Press.

Featherstone, D. (2012) *Solidarity: Hidden Histories and Geographies of Internationalism*, London: Zed Books.

Featherstone, D. and Griffin, P. (2016) 'Spatial relations, histories from below and the makings of agency reflections on *The Making of the English Working Class* at 50', *Progress in Human Geography*, 40(3): 375–393.

Foster, J. (1990) 'Strike action and working-class politics on Clydeside (1914–1919)', *International Review of Social History*, 35(1): 33–70.

Gillan, K., Pickerill, J. and Webster, F. (2008) *Anti-War Activism: New Media and Protest and the Information Age*, Basingstoke: Palgrave Macmillan.

Gray, N. (2015) *Neoliberal Urbanism and Spatial Composition in Recessionary Glasgow*, Unpublished PhD thesis, University of Glasgow. Available from: http://theses.gla.ac.uk/6833/1/2015GrayPhD.pdf (accessed 12 May 2017).

Griffin, P. (2015a) 'Labour struggles and the formation of demands: The spatial politics of Red Clydeside', *Geoforum*, 62: 121–130.

Griffin, P. (2015b) *The Spatial Politics of Red Clydeside: Historical Labour Geographies and Radical Connections*, Unpublished PhD thesis, University of Glasgow. Available from: http://theses.gla.ac.uk/6583/1/2015griffinphd.pdf (12 May 2017).

Harvey, D., Brace, C. and Bailey, A. (2007) 'Parading the Cornish subject: Methodist sunday schools in west Cornwall, c. 1830–1930', *Journal of Historical Geography*, 33(1): 24–44.

Heath, N. (no date) *Anarchists against World War One: Two Little Known Events–Abertillery and Stockport*. [Online]. Available from: www.katesharpleylibrary.net/p8d0px (accessed 17 December 2015).

Kenefick, W. (2007) *Red Scotland: The Rise and Fall of the Radical Left, c. 1872 to 1932*, Edinburgh: Edinburgh University Press.

Knox, W. (1987) *James Maxton*, Manchester: Manchester University Press.

McLean, I. (1983) *The Legend of Red Clydeside*, Edinburgh: John Donald.

Melling, J. (1983) *Rent Strikes: People's Struggle for Housing in West Scotland 1890–1916*, Edinburgh: Polygon Books.

Milton, N. (1973) *John Maclean*, London: Pluto Press.

Mueller, J. (1991) 'Changing attitudes towards War: The impact of the First World War', *British Journal of Political Science*, 21: 1–28.

Navickas, K. (2009) 'Moors, fields, and popular protest in South Lancashire and the west riding of Yorkshire, 1800–1848', *Northern History*, 46(1): 93–111.

O'Connor, E. (2004) *Reds and the Green: Ireland, Russia and the Communist Internationals 1919–43*, Dublin: University College Dublin Press.

Pearce, C. (2001) *Comrades in Conscience: The Story of an English Community's Opposition to the Great War*, London: Francis Boutle.

Pennybacker, S. (2009) *From Scottsboro to Munich: Race and Political Culture in 1930s Britain*, Princeton, NJ: Princeton University.

Purvis, P. (2007) 'The Pankhursts and the Great War', in A. Fell and I. Sharp (eds.), *The women's movement in wartime: international perspectives 1914-1919*, Basingstoke: Palgrave Macmillan, 141–157.

Royle, T. (2006) *The Flowers of the Forest*, Edinburgh: Birlinn.

Schmidt, M. and van der Walt, L. (2009) *Black Flame*, Edinburgh: AK Press.

Smyth, J (1992) 'Rents, peace, votes: Working-class women and political activity in the First World War', in E. Breitenbach and E. Gordon (eds.), *Out of Bounds: Women in Scottish Society 1800–1945*, Edinburgh: Edinburgh University Press, 174–196.

4 The First World War in Palestine

Biographies and memoirs of Muslims, Jews, and Christians

Eyal Berelovich and Ruth Kark

Introduction

The First World War, generally regarded as a watershed in world history, was an important turning point in the history of Palestine.[1] Over the course of the four years of the war, battles took place in Palestine and the Sinai Desert between the armies of the Ottoman and British Empires. Extensive geographic-historic and historical research has examined the war through the eyes of Palestine's local inhabitants, most of which focused on how communities or individuals coped with the war and on the reciprocal relations within and between the different religious groupings (Berelovich 2011: 7–12). Research on the Home Front also dealt with disease, famine, locusts, ethnic violence and population transfer, relief workers, and civilian organizations (Efrati 1991: 16; Eliav 1991: 11–13; Tamari 2011; Çiçek 2014: 80–81; Halevi 2014: 40–51).

There is, however, a void in the research with regard to what the local inhabitants knew about the war and to what extent they identified with the Ottoman cause. This chapter will examine first, how the Home Front accessed news, and the flow of information to learn about the military situation at the front, as well as the specific situation in Palestine itself. Second, we will look at how the inhabitants perceived the war; and third, whether the war changed the perception of the inhabitants, as individuals and as communities, on the Ottoman Empire and on Palestine and its possible post-war future. To do so, we use diaries and autobiographies of men who were Arab Muslims, Arab Christians, local Jews, and foreign Christians, all of whom lived in Palestine during that period.

We focus on five individuals who depicted their daily life in Palestine in diaries during the First World War from its beginning to the British conquest of Jerusalem (9 December 1917) and several who wrote memoirs.[2] They represent the diversity of the local population in Palestine. Khalil al Sakakini (1878–1953) was an Arab Orthodox Christian from Jerusalem, a renowned teacher and school principal (Sakakini 2004). Ihsan Turjeman (1891–1917), also from Jerusalem, was an Arab Muslim soldier in the Ottoman Army, who served as a clerk in the 4th Army logistical center in the city (Tamari 2011). Mordechai Ben Hillel Hacohen (1856–1936) was a Jewish businessman, journalist, and Zionist activist (Hacohen 1929). Shmuel Yehudai (1892–1964) was a Jewish laborer in the Jewish agricultural settlements (Yehudai n.d.). Antonio de la Cierva y Lewita, also known as Conde de Ballobar (1885–1971), was the Spanish consul in Jerusalem and the only diplomat who served the entire length of the war in Palestine (Ballobar 2011). We also cite memoirs relating to the period. Among them are those of Lars Lind (1891–1981) a Swedish national and member of the American/Swedish Colony (a Christian commune) in Jerusalem from 1896, who worked during the war as a professional photographer; Rachel Yanait Ben Zvi (1886–1979), a Zionist activist who travelled inside Palestine during the war; and Wasif Jawhariyyeh (1874–1972), an Arab *'uod* musician from Jerusalem who spent part of the war in Jericho (Lind, unpubl. ms.; Yanait Ben Zvi 1969; Lind and Wallström 1981; Jawhariyyeh 2002).

Diaries and autobiographies as an historical source have been examined in detail by a number of researchers. Abigail Jacobson compared diaries written by two Arab Jerusalemites during the war, mainly in the context of the "multiplicity of identities and the process of surrounding negotiation among Ottomanism, Arabism, and local identities". She viewed diaries not merely as testimonies of individuals, but also as shedding light on larger social groups, reflecting a social process and environment (Jacobson 2011).

In the context of the First World War in the Palestine, Salim Tamari, who wrote about several local diaries recorded during the War argued,

> The power of wartime diaries lies in their exposure of the texture of daily life, long buried in the political rhetoric of nationalist discourse, and in their restoration of a world that has been hidden by subsequent denigration of the Ottoman past – the life of communitarian alleys, obliterated neighborhoods, heated political debates projecting possibilities that no longer exist, and the voices of street actors silenced by elite memoirs: soldiers, peddlers, prostitutes, and vagabonds.
>
> (Tamari 2011: 4–5)

Diaries, therefore, express both the writer's own experience and memory, and the collective experience and memory of his or her family and community, as Jay Winter and Emanuel Sivan argued in their seminal 1999 book: "collective remembrance is public recollection. [. . .] when people enter the public domain [. . .] they bring with them images and gestures derived from their broader social experience" (Winter and Sivan 1999: 6).

However, we have to keep in mind the pros and cons of using diaries and memoirs as a historical source. Their historical importance is vested in their immediacy – the fact that they were written as the events unfolded. They also reflect the inner experience and feelings of the individual who wrote them, and thus are personal and subjective. In some diaries we can identify disclosure or self-censorship, motivated by fear that the diary will be exposed, of certain topics that might endanger the writer or others. Sometimes diaries do not relate to the public domain at all (i.e., the Second World War diary of Anne Frank). Memoirs may be problematic in that they were written after the events took place and are dependent on memories that are not always accurate. Esther Farbstein, writing about the Holocaust, considered diaries versus memoirs as historical sources. She mentions the importance of the motivation behind writing diaries and memoirs, and that some of the writers thought about future readers of their documentation. There was a tendency to embellish and amplify when writing memoirs. She examined the reliability and validity in one case study using a diary and memoir written by the same person. There was extensive concordance. However, she found the diary to be more informative whereas the memoir was more detailed, descriptive and emotional (Farbstein 2002: i–xi, 549–73).

We argue that the historical context is the key to converting the individual experience to a collective one. The historical events define the general framework for the individual's own experience, and redefine the reciprocal relations between the individual and their family, the ethnic and religious community, and nation. The significance of the events the writer is living through and depicting in his diary cannot be understood without understanding the relations between the writer and their community. Thus, before we begin our discussion on the war, we will examine the local population in Palestine before the war by reviewing the local administration, infrastructure, and social structure.

Administration and infrastructure

Before the war, Palestine was not a single administrative unit but three provinces (*sancak*) not divided on a geographical basis. The district administrations were subject to major reforms in 1864, 1871, and 1913 in which the central government gave the local population more say in the way their districts were governed and in future planning ('Aud 1969: 62–138; Gerber 1985: 122–23; Findley 1986: 5–7).

The transportation network was not directly connected with the center of the Ottoman Empire, due to a lack of funding, and the inability to dig through the Taurus and Amanos mountains. Furthermore, the government did not draw up or outline plans for the improvement of the transportation for the Ottoman Empire and for Palestine. In Palestine, the void left by the Ottoman government was filled by foreign private companies, private investors from inside or outside Palestine, and ethnic religious communities (Berelovich 2011). This meant that the main transportation route was via the Mediterranean. The local inhabitants imported raw materials such as coal, cement, and basic food products like flour or coffee through the coastal ports (Avitzur 1972: 49–59). If an effective naval blockade was instituted, the inhabitants would have to live without those raw materials and other necessities.

The social structure

In Ottoman Palestine, individuals navigated daily between Ottoman or foreign nationality. From the Ottoman perspective, the subjects were divided according to their religious affiliation in the *millet* system. Jews, Orthodox Christians, Roman Catholics, and Armenians were given autonomous rights in religious and internal matters such as education or some legal matters. However, from the middle of the nineteenth century, and especially after the citizenship law of 1869 was enacted, every citizen was viewed as a subject of the Sultan regardless of his or her religious affiliation. This law and other reforms eroded the autonomy that non-Muslims had enjoyed under the *millet* system. The ethnic religious communities managed to maintain their educational systems, but had to repel the continuous efforts of the central government to expand its authority over every aspect of the subjects' daily life (Rodrigue 2013: 40–41). According to Aron Rodrigue, one of the outcomes of this process led "to the exacerbation and intensification of ethnic and nationalist sentiment" (Ibid: 41). The local population in Palestine was divided into three ethnic religious communities: Christian, Jewish, and Muslim. The Ottoman government and members of the communities were trying to change the existing social order in contradictory directions.

Tensions between the different ethnic religious communities increased after the local Arab elite and the Jewish residents challenged the Ottoman government with calls for greater autonomy or even independence. Nevertheless, the daily positive interactions between individuals, especially between the local *Sephardic* and Middle Eastern Jews and Muslims and Christian Arabs, were not affected (Chlouche 1931). Adding to the complexity, the ethnic religious communities were internally divided as well. This was especially evident in the Jewish community which, according to Raşid Bey, the governor of the Jerusalem district (1904–1906), was represented by two major groups – the *Ashkenazim* and the *Sephardim* – and further divided into sub-groups according to their place of origin, their nationality, or their ideologies. Furthermore, the different Jewish groups quarreled among themselves and with other communities on various issues (Kushner 1995: 38–40). Division and strife were hardly restricted to the Jewish community, as Raşid Bey's successor, Ali Ekrem Bey (1906–1908), had to deal with an ongoing dispute between the Catholics and the Greek Orthodox over the status quo of Christian holy places in 1907 (Kushner 1995: 142–148). While the government wished to convert the population's identity from ethnic religious to Ottoman, some communities were demanding greater autonomy.

The local population was also divided by ideology as a result of immigration and the import of new ideas. Lastly, the ethnic religious communities faced internal divisions on various issues. Thus, as noted, before the First World War, an individual navigated between five identities: personal, family, ethnic, religious, and Ottoman. As the war continued, individuals had to adapt their identity to the circumstances of that time.

The first attack on the Suez Canal (1914–1915)

At the outbreak of the war in Europe, the Ottoman Empire did not hasten to join either side (Yalman 1930: 68–73; Yasmee 1995: 229–230; Aksakal 2008). Nevertheless, the Minister of War, Enver Paşa, and the Ottoman General Staff began general mobilization, which the local population referred to as *Seferberlik*) (Çiçek

2014: 169–175). Ottoman troops were initially positioned opposite those perceived by the Ottoman General Staff as the main threat to the empire: the Balkan
states and Russia (Başbaknlık Osmanlı Arşivi (BOA): 1–4). In the local Palestine
context, the Ottoman Army at first had no plan for war against British-controlled
Egypt (Türkiye Cumhuriyeti Genelkurmay Başkanlığı 1979: 95). Furthermore,
after the Ottoman Empire entered the war the British military command in Egypt
withdrew its forces from the Sinai Desert and prepared to defend the Suez Canal
from its west bank (Macmunn and Falls, 1928: 1:15).

Before the official declaration of war by the Ottoman authorities, some of the
local inhabitants started to assess the prospect of war in Palestine. Sakakini wrote
about a conversation at the house of Alias Efendi on 18 September 1914 that had
revolved around the Ottoman prospects of occupying British-held Egypt. According
to one speaker, the lack of roads meant that any British force would not have to be
very large in order to repel an Ottoman attack (Sakakini 2004: 2:99, see Figure 4.1)
Apart from contemplating about the future war and the Ottoman military condition,
the local population had to cope with three major changes: first, Palestine became a

Figure 4.1 Sheriff of Medina Preaching 'The Holy War' (*Jihad*) in Medina. Before starting
for Jerusalem (18 November 1914).

Source: Matson (G. Eric and Edith) Photograph Collection, Library of Congress, Washington DC,
LC-DIG-ppmsca-13709–00005.

theater of operations for the Ottoman Army. This meant that the army enhanced its physical presence in Palestine, and its needs superseded those of the local population in every aspect. The local Ottoman government, for example, promptly confiscated resources needed by the army, as well as large buildings. Second, from November 1915, the territory that comprised present-day Syria, Lebanon, Jordan, and Palestine was united under the military and civil authority of the commander of the 4th Army in Palestine, Cemal Paşa. Third, after the Ottomans joined the war, the French and British navies placed a naval blockade on the Ottoman coasts.

According to Lars Lind, the first buildings and resources seized were the "French and British convents, churches, schools and dormitories". The first commodities to disappear were canned goods, imported white flour and sugar (Lind n.d.: 260; see also Ballobar 2011: 32, 36–38). The 4th Army paid in bank notes for food taken from the Jewish agricultural settlements (*moshavot*).[3] Lind's memoir also recalled the unorthodox way the American Colony protected their stock:

> [The American/Swedish colony] had kept only a few pigs in deference to our Moslem neighbors, but now the pen was enlarged. Every scrap of leftover food was saved for the pigs. We figured that while the cows might be commandeered, the Turks would not touch the 'unclean' pigs.
>
> (Lind n.d.: 261)

However, it should be mentioned that it is hard to determine what was paid for in notes or what was taken by force. The official entrance to the war was also accompanied by two declarations: first, the highest Muslim religious authority of the Ottoman Empire declared a Holy War (*Jihad*) (see Figure 4.1, the declaration of Jihad). After the declaration on 18 November 1914, Spanish Consul de Ballobar wrote in his diary:

> By previous announcement and public proclamation, Muslims, and many people who are not, have gathered in the mosque of 'Umar to hear the declaration of Holy War against the Allies. This thing is a little strange because, frankly, if the war is about religion, it should logically be against all Christians.
>
> (Ballobar 2011: 29)

The local Christians were concerned by the Ottoman declaration of Holy War (*Jihad*), as Sakakini wrote on 3 November 1914: "It has been stated, in the Ottoman telegrams, that the Ottoman newspapers, also, see this war as a Holy War. If this is true, then this war has awakened an old spirit and taken us back centuries" (Sakakini 2004: 18)

The second declaration was that all citizens who held French, British, or Russian citizenship had to become Ottoman citizens or face deportation — an order intended primarily to affect the Jewish community (Efrati 1991: 16; Eliav 1991: 11–13; Çiçek 2014: 80–81). Rachel Yanait Ben Zvi recalls in her memoirs that she received a note from Jerusalem on 31 October 1914 calling her to come to the city. There she took part in the Zionist campaign to register as many foreign Jews as possible as Ottoman citizens, as part of the "Ottomanization" process recommended by the Zionist Movement. She wrote, "We viewed the [Ottoman] citizenship as an anchor for rescuing the Jewish settlements" (Yanait Ben Zvi 1969: 343). Some 2700 Jews who were foreign subjects and did not register as Ottoman citizens, were deported to Egypt in December 1914 and January 1915 (Çiçek 2014: 83).

As we see the two Ottoman declarations caused the local non-Muslim inhabitants to redefine their reciprocal relation with the Ottoman Empire. First, there was the religious aspect. Non-Muslim inhabitants, especially the Christians, were declared enemies of the Empire merely because of their religious affiliation. Then, there was the civil aspect. Non-Muslims, especially the Jewish population, many of whom were European subjects, had to decide if they wished to become Ottoman citizens. Thus, at the beginning of the war the Ottoman authorities challenged the Ottoman identity of the local population.

While Hacohen did not mention 'the war' directly in his diary during its first months, he did write about the collateral damage to the local population and the spread of hunger, the locust attack, the increasing food prices, unemployment, arrests by the Turkish authorities, the forced enlistment, confiscations, curfew, and the deportation of Jews. He also discussed the measures that the Jewish leadership took in order to sustain the Jewish community (Hacohen 1929: 1: 1–29). On 12 November 1914, he summarized the speech by the military commander of Jerusalem, Zeki Bey in Jaffa. According to Hacohen, Zeki Bey, who was speaking to the Arab notables of the city, argued, "The inner peace among the different inhabitants is a necessary condition to defeat the external enemy" (Ibid: 29). On 16 December, Hacohen directly encountered 'the war' for the first time: "An English war ship has been seen. [. . .] Much is the commotion in the city, the grocers quickly closed their shops. Fear and panic . . . really? Are the British coming now?" (Ibid: 46).

At the beginning of 1915, the local population began to sense the preparations for the attack on the Suez Canal. De Ballobar, for example, found his attention divided between his various duties, his thirst for news about the war (which was scarce due to a lack of credible information), and travel in Palestine. His efforts to collect information about the Ottoman preparations to attack the Suez Canal drew on rumors, his own observations of the troops marching through Jerusalem, and conversations with Ottoman or German officers (Ballobar 2011: 35). In February 1915, de Ballobar tried to sort through rumors that claimed that the Ottoman 4th Army had crossed the Suez Canal and broken through the British defense lines; or its opposite — that the Ottoman attack was a complete failure (Ibid: 45–51). On 9 March 1915, he wrote that the British may have allowed the Ottomans troops to cross the Canal and then attacked them on the west bank. De Ballobar also spoke to an Armenian infantry soldier who depicted to him the "fog of war" he experienced during the attack (Ibid: 52–53).

By contrast, Hacohen did not actively seek news about the Ottoman attack at Suez. On 8 February 1915, he wrote that he had heard from a Jewish engineer, probably Wilhelm Hecker, who worked for the Ottoman Army (*Davar*, 12 July 1945). According to the news from the telegraph the Ottoman Army had crossed the Suez Canal, broke through the British defenses, and penetrated to Egypt. News of the victory led to celebrations on the streets, and was followed by an official proclamation in Arabic from the commander of Jaffa, Hasan Bey, who called on the Muslims to fight against the infidels. Hacohen interpreted this as a call to fight only against the Jews because: "after all the Christians are the Turks' allies" (Hacohen 1929: 79). Alongside the official telegrams and declarations there were also rumors of a catastrophe. According to Lars Lind, "This debacle for the Turks cost them most of their equipment and heavy guns which had moved up for the showdown (Lind n.d.: 275)", whereas in fact the Ottomans retreated in orderly fashion with most of their equipment. Lind in his memoir also lamented that "[h]ad Germany been able in 1914 or early 1915 to send an army for the capture of the Suez Canal [. . .] the whole history of the war would have been altered" (Wavell 1933: 23–30; Kressenstein 1938: 91–99; Paşa 1973: 154–159; Türkiye Cumhuriyeti Genelkurmay Başkanlığı, *Sina-Filistin Cephesi*, IV ncü Cilt İlk, 203–39). Although he did not confirm the rumors on the failure of the attack until 4 May 1915 (Ballobar 2011: 62–63), de Ballobar wrote on 9 March 1915:

> I am not a military man, but it seems to me a mistake to have attacked the Canal, and instead I think it would have been more effective to have limited themselves to sustaining some force that would have been a constant threat to the English.
>
> (Ibid: 53)

The failure of the attack probably led Ihsan Turjeman on 28 March 1915 to ask, "What will be the fate of Palestine?" (Tamari *Ayam* 2011: 94) On 20 April 1915, he answered prophetically that

> I believe that this war between us and the English and French, as well as the Moscovites [Russians] will last 40 months at the very least. It is true that our army cannot go back to fight in Suez again, after what it saw of British ruthlessness. But I do not think we will see the end of it until the European War is over. We need peace badly. The economic crisis is deepening, and it will not allow us to pursue this war further. Not much is left.
>
> (Tamari 2011: 105)

The failure of the Ottoman attempt to conquer Egypt created a clear distinction between the war front and the Home Front as indicated in an Ottoman map probably made by the Ottoman Army 8th Corps, deployed in Palestine from the beginning of the war until its end (see Figure 4.2, contemporary Ottoman map of Palestine).

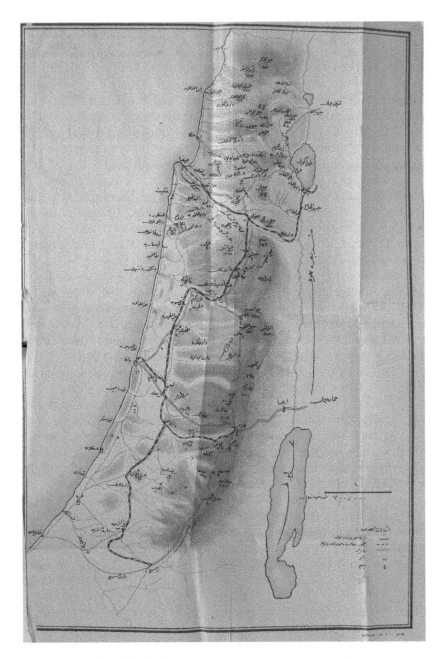

Figure 4.2 1916 map of Palestine.

Source: Unknown author (probably the Ottoman 8th Corps), Filistin Resalesi, (unknown date, probably 1916).

The second attack on the Suez Canal (July–August 1916)

After the failure to break through the British defenses along the Suez Canal, the main attention of both adversaries shifted towards the British and French campaign in Gallipoli (Çankkale). In the Palestine theater of operations, the two sides continued to skirmish on both banks of the Suez Canal, but the British forces were now establishing defense perimeters on the eastern side of the Canal as well.

In 1915–1916, the local population had to deal with five intertwined problems: first, swarms of locusts invaded Palestine in 1915–1916, destroying most of the crops and worsening living conditions. Shmuel Yehudai argued in June 1916 this did more damage to Palestine than the war: "[the locust] is a natural enemy that we can't do much against him [thus] we must look at the disaster and keep quiet" (Yehudai n.d.). (Figure 4.3, contemporary photograph of a plague of locusts). Second, in addition to the locust invasion, there was a drought; and in July fields caught fire during a heat wave (Hacohen 1929: 2: 112–114). Third, because of the Allied naval blockade, the 4th Army was obliged to draw on local natural resources, especially trees or other combustible materials; livestock and other raw materials were confiscated from the local population (e.g. Ibid: 1: 92–94, 1:42, 1: 189–90, 2: 39–40, 2: 79–80). Fourth, the Ottoman paper currency was constantly devalued while prices rose, especially for food (Ibid: 2: 55–56). Hacohen described the situation: "There is hunger in the land, real hunger! Poor the eyes who sees it, poor us who witness it, poor those who were touched by it [. . .]" (Ibid: 1: 215). Although later, on 15 December 1916 he wrote that "Our situation anyway is not so severe and hard as in many countries there [Europe], even in neutral states" (Ibid: 2: 195). Even the army suffered from hunger (Tamari *Ayam* 2011: 171). Fifth, the commander of the 4th Army in Palestine, Cemal Paşa, was trying to convert both Muslims and non-Muslims to identify as Ottoman, rather than with their ethnic religious communities. To do so, he deported the local Jewish and Arab elites, and executed the Arab leaders who had called for reform before the war (Çiçek 2014: 265–269). According to Lind,

> Jamal [*sic*] saw it, the Arabs had first to be re-subjugated and their traitors eliminated before he would take orders from the German General Staff as Enver Pasha [*sic*] was doing in Constantinople. It was this strategic pause which gave hard-pressed Britain time to consolidate her forces in Egypt and on the Canal.
> (Lind n.d.: 269)

The Ottoman civil and military authorities tried to harness the local population to the war effort by publishing newspapers in different languages: The Hebrew *Ha-Herut* (freedom) continued publication during the course of the war. The Ottoman 4th Army also printed its own newspaper (*Musavver Çöl* — The Desert Illustrated) in Beer Sheba (Kushner 2009). They attempted to control the flow of information about the status of the war in the desert and to provide a winning narrative. Nevertheless, because of the daily encounters between Ottoman military personnel and civilians, these attempts were destined to fail.

Figure 4.3 The terrible plague of locusts in Palestine, March-June 1915. Trapping locusts.
Source: Matson (G. Eric and Edith) Photograph Collection, Library of Congress, Washington DC, LC-M32–1433 [P&P].

On top of these distressing factors, there was the war itself. The 4th Army focused on building roads, railroads, and military infrastructure, and tried to assemble and sustain the troops and equipment meant for the second attack on the Suez Canal. Hacohen, looking ahead, asserted that the British conquest would be the only blessing brought by the war: "The bad days will pass, the war will end, the roads and railways will stay and bring fruits and blessings" (Hacohen 1929: 1: 182).

The Ottomans utilized the local population in Palestine, both Jewish and Arab, for the war effort in two ways: first, they turned to the local population to subcontract public works needed for their military infrastructure and logistic support. The local population thus shared the burden of the war as part of the 4th Army's effort to improve the infrastructure in order to increase its ability to rapidly assemble and deploy more troops and accumulate military supplies and resources. Joseph Eliyahu Chlouche (1870–1934), a local Jewish entrepreneur in real estate, industrial, and other business ventures, wrote in his memoir about building the road that connected Latrun and Gaza:

And after several months [from the beginning of the war] the government announced that it will be receiving bids for contracting the road to Gaza via Latrun and Julis, [Meir] Dizengoff and I decided to take this job although we were sure we would barely survive it. It was common knowledge that doing business with the Turkish government will only bleed you dry and exhaust you. However, we were able to release some Jews from the "inferno" that is the Ottoman army.

(Chlouche 1931.; Hacohen 1929: 1: 153–154)

Hacohen wrote on 27 October 1916 that because of the exemption from the military, "many of the Jews and Arabs are willing to work for half pay [. . .]" (Hacohen 1929: 2: 167). The government sued one of the subcontractors, accusing him of issuing work permits to people solely in order to exempt them from military service (Ibid). Second, the Ottoman Army enlisted non-Muslims (Arab Christians, Jews), who formed work battalions (*Tabur 'Ameliyee*). These men were disrespected and received brutal treatment from the Turkish officers and sergeants (Tamari *Ayam* 2011: 126–127).

The accumulation of troops and resources for the second attack on the Suez Canal began after the end of the Dardanelles campaign. De Ballobar mentioned that the Bavarian officer Kress von Kressenstein had been given command of the Ottoman expedition against the Suez Canal, and that German troops would shortly arrive in Palestine in order to take part in the expedition (Ballobar 1996, 2011: 86; Hacohen 1929: 2: 15) (See Figure 4.4 von Kress at Huji). The appearance of the formidable German soldiers (and later, the British) highlighted the stark difference with the poor Ottoman soldier (Ballobar 1996, 2011: 88–92). Sakakini lamented on the poor conditions of the Ottoman soldier on 21 November 1917 in contrast with the well-supplied British troops:

the rain never stopped all of last night, I was in my bed thinking on the condition of the soldiers [. . .] in my thoughts I went from the Ottoman Army to the English army and imagined to myself the condition of every one of the armies. I imagine the English soldier wrapped with cotton from head to toe [. . .] then I imagined the Ottoman soldier, hungry, his clothes and shoes are torn, nothing shelters him from the rain.

(Sakakini 2004: 123)

In 1916, the local population received official news of success on other fronts and in the Sinai Desert, followed by official or unofficial celebration in the army and outside of it (Ballobar 1996, 2011: 86; Hacohen 1929: 2: 75). Wasif Jawhariyyeh recalled that after the second battle for the Suez Canal,

An official telegram came about the Turkish Army victory in Egypt, with the help of its German ally. They seized the Suez Canal from the British and the Ottoman Empire had freed it. This news had spread across the Empire and its inhabitants celebrated in nationalist festivals in city mosques across the Ottoman Kingdom.

(Jawhariyyeh 2002: 227)

However, the claim of an Ottoman victory was false.

Von Kress and Col. Gott at Huj.

Figure 4.4 Kress von Kressenstein and Col. Gott at Huj, 1916.

Source: 1916. Matson (G. Eric and Edith) Photograph Collection, Library of Congress, Washington DC, LC-DIG-ppmsca-13709–00098.

The unsuccessful second attack on the Suez Canal marked the end of the Ottoman campaign in the Sinai Desert and the beginning of the battle for Palestine. The success of the British forces in the Sinai Desert renewed a strategic question in Whitehall: could Britain win the war by first defeating the Ottoman Empire? (Hankey 1961: 636–637).

The First Battle of Gaza (26 March 1917)

The advance of the British forces through the Sinai Desert powerfully affected the local population, manifested in thoughts about the future of Palestine and the fate of its population. Nevertheless, representatives from the different sectors of the local population in Jerusalem still volunteered to promote the Ottoman war effort, for example by organizing in 1917 a bazaar in aid of the Red Crescent (see Figure 4.5, Red Crescent Bazaar, 1917) The writers were hearing, seeing, and feeling the war itself for the first time. On 3 January 1917, de Ballobar wrote,

> No news of the war, or rather of the peace. A person told me the day before, "Until the British take control of this country they will not make peace,

Figure 4.5 Bazaar held at Notre Dame de France in aid of the Red Crescent Society, Jerusalem, 1917.

Source: American Colony Photo Department, Library of Congress, Washington DC, LC-DIG-matpc-08170.

because they need it for the defense of the Suez Canal. But they better hurry, since the Amanus tunnels will be finished in a few months, and then it will be easy to bring in various German divisions."

(Ballobar 2011: 126)

Thus the advance of the British forces through the Sinai Desert led to new questions and new hopes, but also anxiety, among the local population.

On 10 January 1917, de Ballobar spoke to the German consul in Jerusalem, Dr. Johan Brode, who told him that the British forces had not moved from el 'Arish, but other German officials told him that 50,000 British soldiers were indeed moving towards Palestine and could occupy Jerusalem within a couple of weeks. According to de Ballobar, the Ottoman soldiers then marching towards Beer Sheba and Gaza: "are convinced of the near and miserable ill fortune that awaits them and they have decided to spend the little money they still have on something to eat" (Ibid: 127). He also complained about the increase in food prices, the British air raids, the French naval bombardment of Jaffa (4 March 1917), and the expulsion of some Jewish leaders from Jerusalem to Damascus (Ibid: 126–49).

Hacohen noted in his diary entry for 28 December 1917 that the government was ceasing its infrastructure projects in the desert, and ordering its troops to concentrate near Beer Sheba. The British Army, he wrote, was moving through the desert in great numbers. At the end of the entry he asked, "Will the English be upon us from the desert and not from the sea?" (Hacohen 1929: 3: 6). On 12 January 1917, Hacohen wrote about the effect war rumors had in Tel Aviv:

There is room for the street strategists and politicians of Tel Aviv to display their knowledge and assumptions. According to the news and opinions on the street, that can't be verified, the English already advanced to Gaza. And [*sic*] the Turks are preparing to leave Beer Sheba. And there are rumors that the Turkish government ordered the population of Gaza to leave the city, and the government will do so to every city when the enemy will approach it.

(Ibid: 26)

These rumors materialized. At the beginning of March 1917, all of the 40,000 residents of Gaza, overwhelmingly Arab Muslims, were ordered by the Ottomans to evacuate the city in 2 days. They left and moved to near and far away villages and towns in Palestine and Syria. The city was ruined during the conflict and there were long-term ramifications due to its slow pace of recovery and reconstruction (Ibid: 103).

On 28 March 1917, Hacohen with other residents of Jaffa and Tel Aviv received the deportation order. The deportation order excluded farmers, winery workers, field owners, and the teachers and students in the *Mikve Israel* Jewish School of agriculture. He also mentioned that the residents could go wherever they wished, except Haifa or Jerusalem. On 2 April 1917, Hacohen wrote: "There is no Jaffa" (Hacohen 1929: 120; Halevi 2014: 41–51).

Rachel Yanait Ben Zvi did not mention the first battle of Gaza in her autobiography, focusing mainly on the deportation of the Jewish population from Tel Aviv:

> We are imagining hearing cannon sounds from the southern front. The anxiety is deepening [. . .]. They say that the government is conspiring to declare a general deportation from Tel Aviv. The ground is beginning to burn beneath our feet [. . .] the danger of deportation is looming above the Jewish settlements like the sword of Damocles. Until now, the Ottomans whispered about the deportation but now they are saying it out loud [. . .] deportation means destruction.
>
> (Ben Zvi 1969: 463)

The deportation of the local population from Jaffa and Tel Aviv was the greatest trauma for the Zionist Jewish population, which feared that this meant the end of Zionist settlement in Palestine. A short-term effect of the deportation was that some Jewish settlements were flooded with refugees. The lack of resources led to tension between the refugees and those who received them in the settlements.

On 29 March 1917, de Ballobar wrote first, about the official telegram that informed the population about the Ottoman victory in Gaza. Second, he received a commission from the Spanish colony of Jaffa, informing him that they were to evacuate from Jaffa in 12 days. He later (30 March 1917) went to speak to Cemal Paşa about this order. Cemal Paşa told him that all Ottoman and neutral inhabitants would be evacuated north via rail and at the government expense (Ballobar 2011: 47–48). The end of the first battle of Gaza left the local population in a state of uncertainty about their own present and future.

The second battle of Gaza (17–19 April 1917)

On 20 April 1917, Hacohen, who resided in Haifa, heard the sound of artillery from the south (Hacohen 1929: 3: 158). One day later, he was worried about his son David who had enlisted in the Ottoman Army:

> But for whom and for what are our youth going to work in Turkey's war army? Inasmuch as we all recognize the decay in the Turkish regime [. . .] we are all raising our eyes day and night to the English, because they will come and conquer the land, and we are willing to accept with open arms every army that will come and redeem us from the Turks.
>
> (Ibid, 159)

On that same day, de Ballobar accompanied the Austrian and German Consuls to Cemal Paşa's office. Cemal Paşa disingenuously informed them of an Ottoman victory in the second battle of Gaza. He also told them that there was a possibility that Jerusalem would be evacuated. On 5 May 1917, however, the intention to order the evacuation of Jerusalem was cancelled (Ballobar 2011: 152, 155).

Following the second battle of Gaza, there was greater uncertainty among the local population in Palestine, especially in the face of contradictory orders issued by the local government to the population. Mainly, there was a sense that British occupation was now unavoidable, and no one could predict what changes that would bring.

The third battle of Gaza (31 October 1917)

At the end of May, Hacohen complained about the slow pace of the British forces. He argued that the British should have landed forces from the sea in Jaffa and Haifa and on every beach in Palestine two and a half years earlier, but for some secret reason they had not done that. "Now they are sluggishly building railroads on the beach and they are supposed to conquer Palestine through the desert and along the coast of the country" (Hacohen 1929: 3: 193). In his opinion: "the British are 'fighting' as they always fought one step forward two steps backwards" (Ibid). On 9 November 1917, Hacohen noted the orderly fashion in which the Ottoman Army was retreating and the fact that they did not loot the towns they passed by. In contrast, Shmuel Yehudai wrote that the Ottoman Army looted and destroyed everything they could (Yehudai n.d.). Resolving this disparity requires further research. On 14 November 1917, Hacohen believed he could feel the British troops and the beginning of the redemption [*Ithalta De'Geula*]. On the following day, the committee of the Jewish agricultural settlement in Rishon Letzion convened to discuss how to welcome the British troops there and in all of the Jewish settlements. Finally, on 20 November 1917 Hacohen returned to Tel Aviv (Hacohen 1929: 3: 93–128).

Between the second and the third battles for Gaza, de Ballobar went on an excursion to Jericho, al-Salt, and Amman, attended several official dinners, and had off-the-record conversations with Cemal Paşa and General Erich von Falkenhayn. Von Falkenhayn, who replaced Cemal Paşa as the military commander of the Palestine campaign in September 1917, explained that the military situation was rather grave because 100,000 British soldiers were advancing towards Jaffa, Hebron, and Jerusalem and German reinforcements had not yet arrived (Ballobar 2011: 164, 167). Only on 8 November 1917 did de Ballobar learn that the British had occupied Beer Sheba and they were threatening Hebron. On 10 November, he wrote that his German and Austrian colleagues were preparing to leave for Damascus because "the danger of occupation [of Jerusalem] was imminent" (Ibid: 169).

The British advance toward Jerusalem and the fighting in the outskirts of the city led Sakakini on 20 November to ask himself

> what I have done in this war [. . .] three years of war had passed and people from east and west are sunk in her dust and burn in her inferno, while I am worried about two cups of milk and a bowl of soup as they say.
>
> (Sakakini 2004: 172)

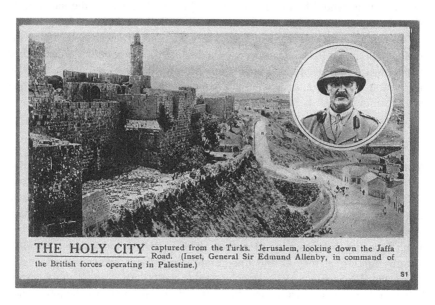

THE HOLY CITY captured from the Turks. Jerusalem, looking down the Jaffa Road. (Inset, General Sir Edmund Allenby, in command of the British forces operating in Palestine.)

S1

Figure 4.6 Jerusalem Captured from the Turks (9 December 1917) with insert of General Edmund Allenby.

Source: Ruth Kark's Postcard Collection, Jerusalem.

On 21 November, de Ballobar wrote about his fears for Jerusalem: "The Turks have blown the rolling stock from the railroad station. I hope to God they don't blow up the mills or the military establishment! I am very much afraid of this because I know the Turks" (Ballobar 2011: 176).

The occupation of Jerusalem, Jaffa and other cities by the British forces marked a stark change in the lives of the local population (see Figure 4.6 Allenby postcard of Jerusalem). Turjeman did not live to see the British occupation of Jerusalem. According to family sources he was murdered by an Ottoman officer in 1917, shortly before Allenby's army entered Jerusalem (Tamari 2011: 60–61, 160). The occupation of Jerusalem marked the end of the war for de Ballobar, Hacohan, and Yehudai. However, it was not the end of the war for Sakakini, who was jailed for helping a Jewish fugitive, and the occupation of Jerusalem continued for nine more months. In that time, the population in British-occupied Palestine tried to cope with the new situation and to establish reciprocal relationships in order to improve their own, and their ethno-religious communities' positions with the new military government. The military situation in that period changed very little. The British forces did not attempt to break through the Ottoman lines in Palestine, but only in Trans-Jordan. The final British breakthrough on 19 September 1918 marked the end of the war in Palestine.

Conclusion

The Turkish General Staff official history argued that the burden of the war effort in Syria and Palestine fell on the Turkish units because of the lack of national sentiment of the local population (Koral 1985: 10, 33). As we have seen, from the beginning of the war the Ottoman government in Syria and Palestine did not try to harness the population to the Empire's war effort, but rather tried to "Ottomanize" it. We argue that a major factor for the lack of national Ottoman sentiment was the implementation of the Ottoman information control policies in Palestine, mainly by spreading half-truths and disinformation through the official communication channels. Because of the short distance between the front and the Home Front, these half-truths could not remain concealed. Furthermore, as we have seen, there were daily encounters between soldiers, serving or defectors, and the local population.

At the beginning, we set out to answer three questions: first, regarding the flow of information and how the Home Front accessed news and learned about the military situation on the Palestine front. For the civilian population news about the war in Palestine were gained from a mixture of official telegrams and proclamations, conversations with an Ottoman or German officer, and through newspaper reports and rumours; as we have seen, the most reliable source of information about the war were rumours from the front. In Palestine during the war reliable information was a rare commodity. The population there was not completely isolated from the Ottoman Empire or from Europe. However, the lack of reliable information from the Ottoman authorities led to another rift between the local population and the government.

The second question looked at how the inhabitants perceived the war, and how perceptions changed as the war went on. At the beginning, it was defined by the framework of their daily lives. As the war progressed and hardships increased, the local population viewed it as an individual war for survival, and at the end of the war as an opportunity for salvation from the yoke of Ottoman rule. No one knew what to expect from the new rulers. How would the British divide the occupied territory? And with which ethno-religious group would they prefer to work in designing the future of Palestine? Although they lacked concrete information, individuals and ethno-religious groups nevertheless began to plan their futures.

Third, we asked whether the war changed the perception of the inhabitants, as individuals and as communities, of the Ottomans and of Palestine itself and its future. Initially, Ottoman rule over Palestine was not challenged by the Arab majority of the population, although we see an exception in the writings of Ihsan Turjeman. He, and several other Muslim intellectuals with local Arab identity, did not predict the fall of Palestine at the hands of the British forces, and was uncertain whether Ottoman rule would persist or be replaced. The Jewish community in Palestine tried to save the Jewish sector in the cities and agricultural settlements by urging Jews to become Ottoman subjects and serve in the Ottoman Army. Ultimately, however, as a result of hunger, epidemic diseases, and the actions of the local Ottoman government, the people of Palestine contributed little to the war effort, focusing their efforts on survival. They withdrew into their own nuclear families and their ethnic religious communities. Thus, when the British came to occupy Palestine, they encountered a fragmented society in which each of its ethnic religious communities imagined a different future for Palestine.

The end of the First World War and the 1916 Sykes-Picot agreement between the United Kingdom and France created a new map for the Middle East, one that is now crumbling a century later. In other words, the mandate created a new 'national-geographical' paradigm to differentiate the local population. The new division of the Middle East allowed the British and French government to solidify their own interests; however, it also forced the local population to redefine themselves again and again since then.

The ramifications of the First World, the Sykes-Picot agreement and colonial heritage war in the Middle East and Palestine remain evident in the current state of conflict, a century later (Muir 2016).

Notes

1 In this article, the boundaries of Palestine coincide with the borders of Palestine under the British Mandate.
2 We do this because after stabilization of the front line in Palestine in December 1917 the historical context is changed.
3 (Central Zionist Archives [CZA], Record Groups/Files: L2/59–10 and L2/59–16; J85/86 11; L2/59–17; S15/20165–4–8, 10, 14, 15, 19–20, 27, 36, 41, 48–50, and 77).

Bibliography

Aksakal, M. (2008) *The Ottoman Road to War in 1914*, Cambridge.
'Aud', 'Abed el 'Aziz (1969) El Idara el Othmaniya fi Wilayet Suriya, Cairo.
Avitzur, S. (1972) *Nemal Yaffo*, Tel Aviv.
Ballobar, C. de [Antonio de la Cierva y Lewita] (1996, 2011) *Jerusalem in World War I: The Palestine Diary of a European Diplomat*, edited by Eduardo Nanzano Moreno and Roberto Mazza, London.
Başbaknlık Osmanlı Arşivi (BOA), İ. Duit.00055,*1329 Taşkilat Umumiye-i Askeriye nizamnası*, İstanbul: Matba-i Askeriyye, 1–4.
Ben-Zvi, R.Y. (1969) *We Ascend*, Tel Aviv.
Berelovich, E. (2011) *Eretz Israel ve-ha-Milhama*, Unpublished M.A. Thesis, The Hebrew University of Jerusalem.
Central Zionist Archive (CZA), Jerusalem, Record Groups/Files: L2/59–10 and L2/59–16; J85/86 11; L2/59–17; S15/20165–4–8, 10, 14, 15,19–20, 27, 36, 41, 48–50 and 77.
Chlouche, Y.E. (1931) *Parashat Hayai*, Tel Aviv. http://benyehuda.org/chelouche/parashat_xayay.html (accessed 12 May 2017).
Çiçek, M.T. (2014) War and State Formation in Syria, London.
Efrati, N. (1991) *The Jewish Community in Eretz-Israel during World War I*, in Hebrew, Jerusalem.
Eliav, M. (1991) *Siege and Distress: Eretz Israel during the First World War*, in Hebrew, Jerusalem.

Farbstein, E. (2002) *Hidden in Thunder: Perspectives on Faith, Theology and Leadership during the Holocaust*, in Hebrew, Jerusalem: Mossad Harav Kook.

Findley, C.V. (1986) 'The evolution of the system of provincial administration as viewed from the center', in D. Kushner (ed.), *Palestine in the Late Ottoman Period: Political, Social and Economic Transformation*, Jerusalem, 2–29.

Gerber, H. (1985) *The Ottoman Rule in Jerusalem*, Berlin.

Hacohen, Mordehai Ben Hillel (1929) *Milhemet ha'amin*, 4 vols, Jerusalem.

Halevi, D. (2014) 'Gaza and its people: Exile and ruin', *Zamnim*, 126: 40–51.

Hankey, M. (1961) *The Supreme Command*, Vol. 2, London: George Allen & Unwin.

Jacobson, A. (2011) *From Empire to Empire: Jerusalem between Ottoman and British Rule*, Syracuse, NY.

Jawhariyyeh, W. (2002) *Al Quds el Othmaniyeh fi Mudhakarat al Jawhariyyeh*, 3 vols, edited by S. Tamari and I. Nasser, Jerusalem.

Koral, Necmi (ed.) (1985) Birinci Dünya Harbi İdari Faaliyetler ve Lojistik, Vol. 10, Ankara.

Kressenstein, K. von (1938) *Mit den Türken zum Suezkanal*, Berlin.

Kushner, D. (1995) *A Governor in Jerusalem*, Jerusalem.

Kushner, D. (2009) 'Musavver Çöl, an Ottoman magazine in Beersheba towards the end of World War I. Hebrew', *Cathedra*, 132(June): 131–149.

Lind, L.E. (n.d.) *Memoir*, handwritten manuscript.

Lind, L.E. and Wallström, T. (1981) *Jerusalems Fararna*, n.p.

Macmunn, G. and Falls, C. (1928) *Military Operations Egypt and Palestine: From the Outbreak of War with Germany to June 1917*, Vol. 1, London.

Muir, J. (2016) 'Sykes-Picot: The map that spawned a century of resentment', *BBC News*, 16 May, Irbil. Available from: www.bbc.com/news/world-middle-east-36300224.

Paşa, D. (1973) *Djemal Pasha, Memories of a Turkish Statesman 1913–1917*, New York.

Rodrigue, A. (2013) 'Reflections on millets and minorities', in R. Kastoryano (ed.), *Turkey between Nationalism and Globalization*, London, 36–47.

Sakakini, K. (2004) *Yawmiyyat Khalil Sakakini*, Vol. 2, Jerusalem.

Tamari, S. (2011) *Year of the Locust: A Soldier's Diary and the Erasure of Palestine's Ottoman Past*, Berkeley, CA.

Türkiye Cumhuriyeti Genelkurmay Başkanlığı (1979) *Sina-Filistin Cephesi*, IV ncü Cilt İlk, Ankara.

Wavell, A.P. (1933) *The Palestine Campaign*, London.

Winter, J. and Sivan, E. (1999) 'Setting the framework', in J. Winter and E. Sivan (eds.), *War and Remembrance in the Twentieth Century*, Cambridge, UK, 6–40.

Yalman, A.E. (1930) *Turkey in the World War*, New Haven, CT.

Yasmee, F.A.K. (1995) 'Ottoman empire', in K. Wilsoin (ed.), *Decisions of War 1914*, London, 229–267.

Yehudai, S. (n.d.) Unpublished handwritten diary. Transcribed by Moshe Yehudai. Courtesy of Rabbi Moshe Yehudai, February 2015.

5   Malta in the First World War

An appraisal through cartography and local newspapers

John A. Schembri, Ritienne Gauci,
Stefano Furlani and Raphael Mizzi

This chapter explores the experience of Malta during the First World War. While the island saw no direct hostilities, the war affected Malta in the form of services rendered to the British Empire and to the Allied war effort due to its strategic location. The chapter explores the historical geography of Malta's involvement in the First World War through the analysis of three maps and illustrated with contemporary images to present the geographies of Malta as its land uses were affected by the war. Details of hospital sites, capacity, and main function together with the upgrading of vital roads, at a scale of 1:175,000 in Sir W.G. Macpherson's *History of the Great War* (1921) are utilised. By using the map and information from other sources, and the resource Google Maps Engine GIS-Lite, an interactive web-based database has been developed, which will provide users with possibilities to explore Malta's overlooked contribution to the Great War effort as has been developed that will provide users with interactive possibilities to explore information pertaining to Malta's contribution to the Great War effort. Two other maps present a detailed comparative study of 2-inch scale maps (1:31,680) of Malta produced in the early decades of the twentieth century: in 1910 (before the conflict) and 1923 (after the war). The differences in land-use development demonstrate the public, military and other works performed on land and on the coast in order to enhance the island's services both during the war and in its immediate aftermath. These services can be divided into three: the northern hospital facilities, the southern-based new fuel installation depot; and, the seaplane base at the southern extremity. The changes that affected the main harbours, in order to accommodate wartime needs, will also be reviewed. In addition, a contemporary newspaper study revealed interesting elements of how the war affected the local population.

Introduction

This study about the role of Malta and the Mediterranean more broadly is justified for a number of reasons. First, the chapter draws attention away from the usual sites of the Western Front that were the main foci of schemes of military strategy and conflict. We therefore highlight the importance of the spaces beyond the 'front line', where the Mediterranean was considered a backwater and marginal to the hostilities, but which ultimately had a strong bearing to the outcome of the war.

Furthermore, the hostilities in Greece and Turkey provided a different topographic challenge with their rugged and mountainous terrain together with a much warmer climate, especially in summer. Armies had to use different provisioning and supply strategies, especially by having to change the mode of transportation from land to sea to land again, to embark and disembark supplies and munitions, casualties, troops and other servicemen and women. It must be recognised that part of the global nature of the conflict is connected to the way in which the successful prosecution of the war depended on an ability to control seas and move material and personnel efficiently across vast spaces. These include successful movements within similar semi-enclosed seas (such as the Baltic, the Black Sea and the Persian Gulf) thus controlling the Mediterranean played a crucial role. Sitting at a strategic point between the Adriatic, Aegean, Levant and Western Mediterranean, Malta played a significant – though largely overlooked – role within the conflict. The war also spread to islands and peninsulas that were initially considered as peripheral to the conflict but gained in importance as the hostilities increased. Footholds such as Gibraltar, Malta, Crete, some Aegean Islands and the Crimea became important locations within the (so-called) periphery. The seas surrounding these islands and the channels and waterways leading to them, extended the sites of conflict on the marine environment, into an important role. In this context, the war also involved surface-to-surface maritime hostilities and deep-sea surveillance for mines and submarines. It essentially became a volumetric war with the introduction of air reconnaissance to monitor surface and sub-surface activities in addition to land-based undertakings. Amphibian landings were also important, as witnessed in Gallipoli. In addition, the nature of the conflict owed its global spread to the imperial connections with the number of colonies possessed by the major belligerents. These included the multicultural nature of the British Empire and the African colonies under the Axis powers. But it was in replicating Britain's Home Front as a military and hospital base that Malta came to find as its principal task.

The geopolitical scenario

The Maltese Islands that include Malta, Gozo and Comino, have a long history of occupation–generally by the powers that either controlled the central Mediterranean or were influential in the broader political events that shaped Europe (Blouet 1992). Formally coming under full British control in 1814, Malta played an important role in the maintenance of what was described as a *Pax Britannica*, as a major base through which Mediterranean waterways were kept open for trade and naval traffic.[1] The geography of Malta lent itself to being an efficient base in the central Mediterranean that was somewhat different to other British possessions. Its excellent harbours were classified as 'all-weather', and a substantial skilled population lived in close proximity to shipyards well versed in the technical needs of ship repair and servicing. Malta also had a distinct medical history spanning at least three centuries that was to provide a valuable role to the British war effort.[2] Having been under British rule for a century, the knowledge of English was good enough for efficient communication, both at the technical and social levels. These initial contacts with the British gave the Maltese an increased sense of purpose with the British fleet using the Maltese harbours and utilising the same fortifications built by the Knights. However, as the technology of war advanced and coal-fired steam engines replaced wind-powered sails, the local victual services also underwent rapid changes, which can be set in context of an increasingly perilous global geopolitical situation.

At the start of the twentieth century, European capital cities were at the 'height of their global reach' (Emmerson 2013), with leading states positioning themselves for territorial, maritime and diplomatic supremacy. Thus the main political powers sought to gain as much control of the Mediterranean as possible (Frendo 2012). The establishment of the Triple Entente between the United Kingdom, the Russian Empire and the Third French Republic aimed to safeguard their interests against the Triple Alliance formed between Germany, Austria-Hungary and Italy. The Mediterranean thus featured in its perennial diplomatic fissures through the different allegiances, the Russian need for a warm seaport, and the slow decline of the Turkish Empire (Thomson 1966). As well as developing problems in the Balkans and the fear of social revolution, Britain was also keen to control strategic sites in the Mediterranean to safeguard the connectivity of its global empire.

Although firmly part of the British Empire as a colony with a substantial garrison, modern fortifications and an economic structure based on British military investment and spending, Malta was nonetheless vulnerable to Italian overtures. In addition, the geographic centrality of the small islands (316 km2) in the Mediterranean was strategically enhanced, following the opening of the Suez Canal in the aftermath of the Crimean War. This was imperative to provide a safe area for medical support of British military endeavours, thus enhancing Malta's strategic value, and set the tone for the nature of Malta's contribution during the First World War. Thus, changes to the infrastructure of the harbours on the eastern seaboard of Malta followed the degree of expectant tension in the central Mediterranean. Changes to the Grand Harbour were done mainly to accommodate ship repairing facilities and the construction of two breakwater arms to render the harbour all-weather.

Malta and the First World War

Whilst the Maltese Islands might not have been directly involved in fighting during the war, the islands served the British Empire's war effort in four fundamental ways. As a naval base, with a dockyard solely dedicated to servicing the fleets in the harbours; as a hospital base tending to the sick and injured, with the setting up of hospitals, convalescent homes and camps mainly in the central and northern parts of Malta; as a manpower base, with Maltese personnel serving in the Army, Navy or Labour Corps; and as a seaplane base, with the establishment of an Air Force base at the south-eastern extremity of Malta, at Kalafrana. In addition, a number of roads were upgraded or constructed during the war to enhance communications between the harbours and the rest of the island, military cemeteries were enlarged or newly established outside of intensely built-up areas and the provision of fuel storage in a new depot was developed. Most of these changes were completed during the war for a number of reasons. The first is that, after the first few months, it was clear that the war was going to extend beyond Christmas 1914 – meaning that all the resources necessary had to be employed to counter the prolongation of the war, including the facilities offered by the Maltese harbours. Secondly, the war engulfed Mediterranean sites with the decision to disembark Allied troops at Gallipoli. This momentous decision by the British Admiralty placed the strategic location of Malta nearer to the scenes of the conflict. As a consequence to this situation, German U-boats started to ply around the islands, thus introducing the new technology of underwater warfare to Malta. The result was that the islands had to be equipped with anti-submarine methods: new coastal bases, servicing yards and trained personnel, some of whom would be tasked with spotting submarines from aircraft.

Malta's economy and employment opportunities duly increased and broadened to the extent that 14,000 people were employed in the harbours, out of a population of 212,000 (Blue Book 1910–1911). Indeed, it is estimated that 24,000 Maltese played a direct part by joining the British services during the war. These included a detachment of 864 stevedores and 800 gangers (the head of a gang of labourers working on canals and railways, in this case building pontoons, digging trenches) sent to the Dardanelles and Salonika in the first week of September 1915 (Mackinnon 1916). By the end of the Gallipoli campaign 1,108 men had been dispatched there, and 1,300 to Salonika; 47 with the Royal Malta Artillery; 75 officers in the Services and 300 with the Canadian and Australian Armed forces, a further 325 Maltese were despatched to Salonika in February 1918 through a labour company. Further details are presented in Tables 5.1 and 5.2, which show a breakdown of the personnel into a number of services engaged and in addition 31,739 men voluntarily served the government in Malta (basically, the bulk of the employment apart from agricultural labour). Indeed, with most of the male working population engaged in seasonal physical labour in the fields, most Maltese viewed these opportunities as a means of earning a guaranteed wage all year round. Therefore, to an extent, this alternative 'Home Front' meant that Malta would bear witness to a number of important societal changes as a result of its wartime experience.

Table 5.1 The number of Maltese engaged directly in the war effort.

Service	Number
Coaling	2,400
Sea-going services	1,300
Mine-sweeping	200
Labour parties*	500
Royal Air Force	788
Mercantile marine	630

*Loading/unloading cargo, blasting operations, laying roads, cutting stone and wood, digging wells, stevedore work.

Table 5.2 Selected details of overseas deployment.

Date	Number	Location
1 September 1915	864	Dardanelles
4 and 5 September 1915	800	Salonika
By February 1917	1,108	Gallipoli
By February 1917	1,300	Salonika
By February 1918	325	Salonika

The declaration of war was announced by the Governor of Malta through newspapers. The *Daily Malta Chronicle and Garrison Gazette* (DMC 5 August, 1914) published news under the sections 'European Conflict' and 'Gazetteer of the War' on a daily basis, in addition to listing shipping lost at sea. It also featured 'unofficial despatches from the front', with German casualties reported to raise morale. Furthermore, safeguarding the use of certain goods was controlled by legal means. Certain metals, including pipe lead, ferrochrome, unwrought copper and iron ore were considered 'conditional contraband'. Rubbed and raw or rough tannel hides and skins were also featured on the same lists (DMC, 28 September 1914).[3] Food prices at Valletta market – identified by weight, measure or packed – were published on a regular basis in order to safeguard against unauthorised price hikes (e.g. DMC, 14 November 1914: 9). Another newspaper, *Il-Habbar*, [The Newscaster or The Town Crier] also had a free insert with La Marseillaise and Rule Britannia lyrics with the French and Union Jack flags and captioned as Entente Cordiale 'The Union of Heart and Hand'. Another newspaper, *Il-Habib* [The Friend] had a more serious outlook. Not only were parish priests encouraged to spread its word among their parishioners, it reported the war alongside contemporary poems by local authors.

Some aspects of life, however, seemed to continue as they had before. On 11 May 1916, (pp. 4–5) the DMC reported that the Malta Horticultural Society had affiliated with the Royal Horticultural Society in London. Prizes were announced and results pertaining to successful candidates of the Associated Board of the Royal Academy of Music and of the Royal College of Music were featured in DMC of 9 May 1916 (pp 4–5). A three-page announcement was placed from a newly opened shop at Valletta, requesting clients to have a look at the wide range of options for sale. These included silverware, sports and games goods, electro-plating, cutlery and toys (DMC 21 November 1914). Adverts for auction sales of antiques (DMC, 9 May 1916) were also carried.

The need for fuel was also considered of importance and one-page adverts appeared to focus on offers for the bulk purchasing of fuel for local use. Even though fuel restrictions were in place, many local businesses continued unabated and depot-to-door delivery was recommended. However it was through the advertisements featured in the newspapers that provided a view of the overall local socio-economic situation. A number of strands could be identified. The Peninsular and Oriental Steam Navigation Company had almost daily announcements of its services from Malta to India, China and Australia and 'frequent sailings' to London via Marseilles and Gibraltar. This suggests that maritime travel was still possible even though hostilities ranged in the Mediterranean. One advert occupied a one-third page longitudinal column. Other regular advertisements dealt with the sale of tobacco, cigarettes and spirits. In other words, despite the clear and sometimes restricting wartime situation, many aspects of life continued as before. Data published in the series of Blue Books indicate a decrease in trade with 177 vessels and total of 318,266 tons with 196 vessels and total of 320, 418 tons cleared by Maltese ports in 1917–18. The provenance of the shipping was essentially from Western European, Mediterranean and Far Eastern ports. These figures are markedly less than those reported for 1912–13 with a 2,912 vessels (5.1 million tons) cleared and 2,924 vessels (5.1 million tons) entered from approximately the same ports (National Statistics Office, accessed on 15 September 2016).

The central Mediterranean location of Malta gave it a strong strategic significance for the surveillance of the surrounding seas. In 1904, the potential for Malta as a surveillance station was confirmed by a visiting Royal Engineers Balloon Unit. In 1915 the first successful seaplane flight from Grand Harbour was performed and the coastal indentations and peninsulas in Malta's south-eastern bay were thus earmarked for development. A seaplane base was completed at the Kalafrana inlet in 1916, complete with hangers and spillways and a fuelling depot was developed on *il-Gżira* promontory, between the two inlets of St George's Bay and Pretty Bay. At Kalafrana, 23 seaplanes were assembled from parts shipped over from the United Kingdom, with their main mission being to escort convoys and spot submarines. The need for fuel in Malta was keenly felt at the start of the war and the first consignment of kerosene arrived in 1914.[4]

As the conflict went on towards 1918, businessmen were encouraged to start investing and be "a step before the crowd" (DMC, 6 August 1917, p 5). Adverts for the purchase of typewriters (DMC, 7 August 1917, p 11) became frequent and cosmetic goods came also to the fore with perfumery and creams for women (DMC, 10 August 1917). It was the adverts for cars, however, that particularly stand out in 1918, with car importation and car hire services (DMC, 13 July) together with details for the sale of car accessories, marine motors, gas and oil engines (DMC, 15 July, 1918 p. 3) all being prominent. At this stage the effects of high employment by the local population servicing the war were being felt and at a time of renewed efforts by the Allies to conclude the conflict on the Western Front. As a result, the financial resources acquired by the people permitted shopping for a number of goods otherwise not affordable. The presence of many foreign servicemen and women on military duty and in their convalescent process further enhanced the overall purchasing power present in Malta.

Hospitals in Malta

The strong medical tradition in Malta had a long history before the start of the twentieth century, especially with the arrival of the Knights known officially as the Sovereign Military Hospitaller Order of Saint John of Jerusalem, of Rhodes in 1530. Their centuries of experience and medical knowledge saw them build hospitals to service the needs of the Knights and the local population. This tradition continued throughout the British period, further re-enforced by the experiences of the Crimean War in the 1850s. By the start of the twentieth century, Malta had four military hospitals with 278 beds mainly spread around the densely populated harbours, and one on Gozo. During the early years of the war, a number of hospitals, convalescent depots and homes were built, commissioned and established in requisitioned buildings and schools. These included the Cottonera Hospital, Forts Chambray and St Elmo, the Government School at Sliema, private homes as at Dragonara, Jesuit Colleges such as St. Ignatius and the Governor's Palaces the Auberge de Baviere and San Antonio. Within a couple of years, there were 28 hospitals with over 20,000 beds with the average occupancy at 16,004. In January 1916, there were 334 medical officers, 913 nurses including VADs (Voluntary Aid Detachment), and 2,032 rank and file members of the Royal Army Medical Corps (RAMC) stationed in Malta. Table 5.3 lists details for each hospital in Malta and Fort Chambray in Gozo. The rapid increase in the provision of hospital beds was particularly notable in an eight-month period in 1915, with further expansions noted in the fourth column of Table 5.3, indicating the escalation of the war. Table 5.4 shows details pertaining to Gallipoli and Salonika to provide some contextual background to the hospital expansion in Malta. The hospitals were mostly located in areas around the harbours and the convalescent camps in areas towards the north of the island.

Medical attention was not neglected in the advertisements and announcements in the local press. Vaccination against smallpox for children older than two months was held in every town and village with a schedule detailing the government dispensary place, date and time when the "gratuitous" injection was to be administered. The reason for this was the smallpox outbreak in 1915, which was attributable to arrivals from Tripoli, Tunis and other Mediterranean ports (Cassar 1964). This also indicates that maritime traffic in the Mediterranean was still an ongoing concern. As the war progressed, adverts for soap describing it as a 'huge life saver' (DMC, 3 August 1917 p 2) appeared. Condensed milk was another product advertised. The aim of this was to encourage people to refrain from buying goats' milk from their doorstep directly from the shepherd as this was not pasteurised and gave rise to several diseases. 'Erasmic' [trade name] toothpaste was also placed alongside the condensed milk adverts. Arguably, it can be demonstrated that such a huge state-sponsored investment in medical and hospital facilities, as well as provisions brought into effect a broader expansion in understanding and interest in hygiene and health resources across the islands. The sequence of procedures beforehand began with transferring patients from the injury site to Malta first to a hospital ship then to the Grand Harbour triage station for patients to be sent to the relevant site for medical attention and convalescence.

Alongside the formal activities of the British Imperial state in expanding Maltese medical resources during the war, came the notable blossoming of what

Table 5.3 List of hospitals and convalescent centres in Malta and Gozo with selected histories and data.

Designation	Date of Opening	Initial number of beds	Maximum expansion	Date of closing
Cottonera	Pre-war	167	802	Post-war
Imtarfa	Pre-war	55	1,853	Post-war
Forrest	Pre-war	20	186	Post-war
Valetta	Pre-war	36	524	Post-war
Tigne	2 May 1915	600	1,314	6 January 1919
St. George's	6 May 1915	840	1,412	31 October 1917
St. Andrew's	9 May 1915	1,172	1,258	–
****Dragunara**	14 May 1915	12	20	29 August 1917
Floriana	4 June 1915	600	1,304	30 April 1917
Blue Sisters	6 June 1915	80	120	30 June 1917
Hamrun	8 June 1915	108	117	5 May 1917
***All Saints**	12 June 1915	1,465	2,000	November 1917
Baviere	15 June 1915	105	155	14 August 1917
St. Ignatius	2 July 1915	155	196	
Sisters Hospital, Floriana	10 July 1915	12	31	
St. David's	25 July 1915	464	1,168	1 May 1917
St. Elmo	12 August 1915	218	348	October 1917
St. Patrick's	15 August 1915	1,000	1,168	27 April 1917
***Ghajn Tuffieha**	15 August 1915	2,000	5,000	January 1919
St. Paul's	25 August 1915	240	898	27 April 1917
St. John's	1 September 1915	400	520	9 October 1917
***Fort Chambray, Gozo**	4 October 1915	400	400	13 March 1916
Spinola	6 November 1915	1,000	1,168	27 April 1917
Ricasoli (Bighi)	6 November 1915	800	800	19 February 1916
Manoel	8 December 1915	500	1,184	21 December 1918
****San Antonio**	8 December 1915	50	50	19 March 1916
****Verdala**	9 December 1915	30	30	17 April 1916
***Mellieha**	1 February 1916	1,250	2,000	5 September 1917

*Convalescent depots
**convalescent homes.
Source: Macpherson (1921).

Table 5.4 Number of patients transferred to Malta from Mediterranean theatres of war.

Gallipoli, 1915	Salonika, 1915–16
by 4 May: 1,200 wounded	First week December 1915: 6,341 sick and wounded arrived
by end May: 4000 wounded	
in September 10,000 wounded	January 1916 – casualty numbers fell due to scaling down of Gallipoli campaign
by March 1916 20,000	
At the peak of the Gallipoli campaign, the average number of arrivals was 2,000 per week. By the end of the war in November 1918, Malta hosted 58,000 patients from the Gallipoli campaign and 67,000 from Salonika. A total of 125,000.	But Summer 1916 increase of casualties due to Salonika campaign
	By April 1917, the number of hospital beds was 25,522.

Source: Laferla (1947).

today would be called the 'third sector' services. These included also the request by the British authorities to enlist the assistance of the Civilian Government's clinical bacteriological work, since the Service's laboratories could not cope with the demand (Ellul-Micallef 2013). Whilst hospital facilities were relatively evident as new large-scale building projects, it is easy to overlook the concomitant expansion of civil society organisations. There are numerous notices in the DMC and other newspapers connected with the growth of organizations involved in assisting the wounded in their rehabilitation. St John Ambulance Brigade, The British Red Cross, the Maltese Boy Scouts, Young Men's Christian Association and various groups from towns and villages all participated in these efforts. Indeed, what could have been considered a first in Christian Ecumenism within Malta were the joint efforts made by the Catholic Church, the Church of England and the Church of Scotland. Activities ranged from tea parties, outings, picnics and sports and games for the wounded and convalescent. The provision of medical services coupled to benevolent activities aimed at providing assistance, together with a measure of entertainment and physical activities, united the religious, medical, sporting and philanthropic organizations in Malta into one main unified scope.

Cartographic representation

The geography of World War I in Malta can be enhanced by examining a number of maps that reveal the distribution of hospitals, beyond detailing their location and main specialization. The main hospitals were grouped around the main harbours, thus decreasing the distance between the triage station and the primary medical facilities. The road pattern then facilitated the distribution of the patients throughout the rest of Malta, so that the convalescent camps were located away from the main hospitals, in more secluded coastal areas. A modern industrial war required a similarly modern and efficient 'industrial' style organisation of medical facilities and transportation links. Thus, the existing road network was improved and utilised in the transport of patients and their carers, food and medicines. Moreover, the increased activity in the harbours became more efficient through the transportation of workers using the railway line that passed through a number of villages.

This project has created an interactive map, using the application of Google Maps Engine GIS-Lite, which drew on information from a range of contemporary sources, reports and publications.[5] This customised map uses appropriate icons and allows the user to modify the map by drawing, searching and also adding points on to it. This modified data allows a better visualisation by grouping the data in different cartographic layers. This dynamic cartographic representation allows for more specific evaluation of the historical geography of Malta during the First World War.

Several significant changes that affected the harbours and other inland areas in Malta were identified, by comparing two maps published by the Royal Engineers Office, Malta, in 1910 and 1923 (see Figures 5.2 and 5.3). In the Grand Harbour, around which most of the economic and servicing activity took place, the fortifications and marine victualising services were undertaken. A late nineteenth century cluster of changes can be identified in Figure 5.2. This map shows the works done around Grand Harbour in various phases starting with the building of Dock No. 1 a year prior to the opening of the Suez Canal in 1869. Although commissioned before the war, the Canal linked Malta to the Far East and cemented further the island's link to the British Empire. Hence the importance of including dock construction within the harbour, as part of British efforts to strengthen Malta's strategic influence in the central Mediterranean. Additional works followed with a coaling station, more docks and victualing yards and two breakwater arms to the harbour turning it into a more efficient all-weather inlet. The map also indicates the extent of the built-up areas around the harbours and clearly shows the complete erasure of several settlements that had been located on the peninsulas. Malta's naval facilities were among the most modern in the Mediterranean region with the services offered around the harbour complimenting new technologies associated with maritime transport and advanced military strategy, supported by additional services such as fuel storage built away from the main harbours (see Figure 5.1).

Utilising the transport infrastructure of Malta's only train line, a string of hospital facilities – and connected military cemetery – were built inland to the north-east and north-west of Notabile (Mdina) (see Figure 5.3). In another example of how Malta's infrastructure was modernised through the experience of war, dead, dying and injured soldiers could be transported efficiently from the harbour to a range of support facilities, including Malta's general hospital at Rabat.

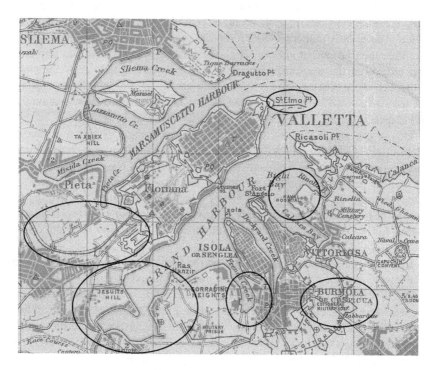

Figure 5.1 A zone in Marsaxlokk Bay with the changes made to the littoral areas of Kalafrana and its seaplane base and the construction of a fuel storage terminal with associated jetty. The base was completely buried in the early 1980s with the construction of a container terminal at the site.

Sources: War Office 1910. Royal Geographical Society (RGS) Collection); War Office 1923, RGS Collection, London.

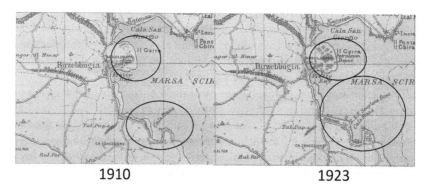

1910   **1923**

Figure 5.2 Sites of Grand Harbour works late nineteenth and early twentieth centuries.
Source: War Office 1910, RGS Collection.

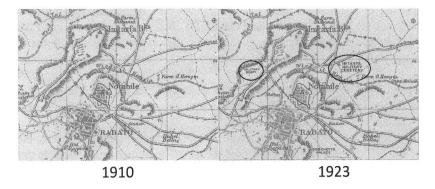

1910   1923

Figure 5.3 Maps of Rabat, Notabile (Mdina) and Mtarfa as at 1910 and 1923.
Source: War Office 1910, RGS Collection.

Conclusion

At the start of the First World War, Britain and its allies endeavoured to halt the advances of Germany and Turkey into the Mediterranean. From being an apparently peripheral island, Malta became a central node within these global endeavours through its medical, military and transport/distribution role. The First World War brought Malta to this strategic forefront for number of reasons. It was appreciably away from airship and balloon attacks; far enough from possible Turkish hostilities, although near enough to engage in modern submarine warfare. It had all-weather deep harbours when compared to Cyprus and was ideally positioned between the UK Naval Dockyards at Plymouth and Portsmouth, and operations in the Middle East. Within Malta itself, and especially within the context of centre-periphery relations, the distribution of medical care, through the location of hospitals, together with the victualing and servicing of naval operations followed a similar core-periphery pattern. Thus, one can discern a common micro-geography of provisioning and supply, care and convalescence, between the military and medical operations. Heavy investment in modern naval facilities was focussed on the deep harbours, with Allied services and fuel depots located at a safe distance and connected with updated and efficient transport links. Likewise, triage and acute medical facilities were located in the urban centres, close to the harbours, and connected to a network of convalescent and long-term care units – and military cemeteries – located through the islands. The latter hospitals made use of the rural and coastal environments as a means to rehabilitate patients.

The scale of the First World War impacted Malta heavily with new employment opportunities to support the many thousands of troops who were passing through the island prior to their deployment in the northeast Mediterranean or Levant. In addition to the number of able-bodied men sent in operations overseas, the building and maintaining of hospitals and erection of hundreds of tents to accommodate the injured provided opportunities for manual work, together with servicing the battleships anchored in the harbours. Whilst previous hostilities that involved Malta saw the island closing its shores and defending itself from the Turks in the 1565 Great Siege and again against the Napoleonic threat and occupation in the last two years of the eighteenth century, Malta opened its shores and harbours in the Crimean War and First World War. The island changed its mission from an introverted defender to an extroverted accomplice of Britain, thus becoming a stronger link in the chain of connective nodes that joined the British Empire.

The impacts of the First World War on Malta can still be felt a century after the event. The depression that followed the end of hostilities caused demographic changes with a decline in the population. In the first instance, there was mass emigration to the United States, with today's Maltese population in the United States tracing, in large part, its origins to this period. Secondly, Malta suffered hugely during the post-war outbreak of the Spanish Flu. On the economic sector; the establishment of Marsaxlokk Bay as an industrial area owes its initial investment to the first two decades of the twentieth century as a site for fuel installations, and the Kalafrana base remained operational up to the mid-twentieth century. At the medical level, the skills imparted to the local population who assisted the British medical personnel servicing the hospitals, continued to enhance the medical traditions in Malta, developed over centuries since the Knights of St John established themselves on the islands. The scale of importance of the war could also be measured through the newspapers publishing continuously the events occurring on various fronts and the nationalism invoked as part of the war effort saw its manifestation immediately after the war. As a result of less British military spending, food shortages and lack of employment civil riots broke out. These were followed by general elections and a representative assembly as was elected. Although an overall lack of political will by Britain to listen adequately to Malta's quest for more political powers could be identified, the Maltese demand for measures of self-government was granted in 1921.

Further impacts can still be seen as commemorative spaces through a number of memorials erected in remembrance of the conflict, the most prominent being the War Memorial and the King George V Merchant Seamen's Memorial Hospital, both at Floriana. Other memorials can be found in Gozo, the King's Own Malta Regiment Memorial, Royal Malta Artillery Memorial and the Air Force Memorial. Other naval memorials can be found at the Msida Yacht Marina and Kalkara Gardens at Floriana. Further memorials are the Naval Cemetery at Kalkara and military cemeteries in Imtarfa, Pembroke and Ta'Braxia (Pieta).[6] Burials also respected the religion of the deceased with Catholics buried at a special section of the Addolorata Cemetery at Paola, Muslims in the Turkish Cemetery at Marsa and other denominations at Pieta Cemetery (Mizzi 1991) Most of the cemeteries are today cared for by the Commonwealth War Graves Commission and a number of them are visited regularly by relatives of the deceased (Zammit 2004). The main war memorial is the site of the annual

November commemoration marking the two twentieth century wars with prominent personalities laying wreaths.

The part played by the Maltese in the war effort is acknowledged by the governor's statement:

> If the hospital arrangements have proved satisfactory, if the lives of 80,000 patients have been made happy during their time in Malta, a great amount of credit is due to the philanthropic work carried out on the island . . . Malta has been given a good opportunity for doing good and she has faced the situation splendidly.
>
> Lord Methuen – Governor of Malta and architect of hospitals expansion programme in Malta. Quoted in Mackinnon.
>
> (1916: 9)

Acknowledgements

Joanna Causon Deguara, Researcher.
David McNeill, Map Librarian Information Services and Resource Division, Royal Geographical Society.
Julie Cole Collections and Enterprise Assistant, Royal Geographical Society.
Jo Dansie, Maps and Manuscripts Reference Service, the British Library.
Ray Polidano, Director-General, Malta Aviation Museum.

Notes

1 For more information on this period of Malta's history, see Fenech (1993), Frendo (2012), Gott (2011), and Holland (2012). Following the occupation of the first Mediterranean possession of Tangier (1682–1684), Gibraltar was declared a British Overseas Territory in 1713. The Mediterranean was of vital importance to British interests especially as a route to the eastern possessions of Egypt, India and in Australasia. With full sovereignty granted over Malta in 1814 the Mediterranean was not far from British interests. The opening of the Suez Canal in 1869, the administrative attention afforded to Cyprus from 1878, and the occupation of Egypt in 1882 paved the way for stronger control of the Mediterranean waterways that were used as routes and channels for British-enhanced control over territories in Asia.
2 The Knights of St John who administered Malta from 1530–1798 had a medical tradition and caring for the sick and wounded was part of their mission. They built hospitals both in their countries where they stayed and on the warfronts.
3 Tannel hides are skins and hides of animals produced as result of kneading animal skin by foot in dung and treating it with chemicals, which makes it less prone to decomposition.
4 It was not until 1920 that a storage depot was built at Birzebbugia with the fuel transferred to Msida and Floriana by mule, a two-hour journey.
5 The map is available and can be accessed on https://drive.google.com/open?id=1gwtoJvIEqpAg_UWJemER0hJNnuw&usp=sharing.
6 Seventy-three Japanese are buried at the Kalkara cemetery. These include 59 who died when their destroyer, Matsu, of the Japanese Royal Navy, was sunk by a German U-boat near Malta on June 11, 1917. Plans to promote Malta among Japanese tourists are at hand.

Bibliography

Azzopardi, A. (1995) *A New Geography of the Maltese Islands*, Malta: Progress Press Co. Ltd.

Blouet, B. (1992) *The Story of Malta*, London: Faber and Faber.

Bonnici, J. and Cassar, M. (1988) *The Malta Railway: Il-Vapur ta'l-Art*, Marsa: Interprint.

Cassar, P. (1964) *Medical History of Malta*, London: Wellcome Historical Medical Library.

Ellul-Micallef, R. (2013) *Zammit of Malta: His Times, Life and Achievements*, Malta: Allied Publications Limited.

Emmerson, C. (2013) *1913: The World before the Great War*, London: The Bodley Head.

Fenech, D. (1993) 'East-West to North-South in the Mediterranean', *Geojournal*, 31(2): 129–140.

Frendo, H. (2012) *Europe and Empire: Culture, Politics and Identity in Malta and the Mediterranean*, Malta: Midsea Books.

Gilbert, M. (2002) *The Routledge Atlas of the First World War*, 3rd edition, Routledge: London.

Gott, R. (2011) *Britain's Empire: Resistance, Repression and Revolt*, Verso: London.

Holland, R. (2012) *Blue-Water Empire: The British in the Mediterranean since 1800*, London: Allen Lane.

Laferla, A.V. (1947) *British Malta*, Vol. 2, Malta: Aquilina.

Mackinnon, A.G. (1916) *Malta: The Nurse of the Mediterranean*, London: Hodder & Stoughton.

Macpherson, W.G. (1921) 'Chapter XIII: The medical services in the Mediterranean Garrison', in *History of the Great War: Medical Services General History based on Official Documents*, Vol. 1, London: HMSO, 235–248.

Map of Malta at 1:175,000 in Sir W.G. Macpherson's *History of the Great War (1921)* Map with hospitals. In Macpherson, W.G (1921) 'Chapter XIII. The Medical Services in the Mediterranean Garrison', in *History of the Great War: Medical Services General History based on Official Documents*, Vol. 1, London: HMSO, 235–248.

Mizzi, J.A. (1991) *Gallipoli: The Malta Connection*, Malta: Technographica Publications.

National Statistics Office, Malta *Blue Book 1910–11*. Available from: https://nso.gov.mt/en/nso/Historical_Statistics/Malta_Blue_Books/Pages/Malta-Blue-Books.aspx (accessed on 15 May 2016).

National Statistics Office, Malta *Blue Book 1917–18*. Available from: https://nso.gov.mt/en/nso/Historical_Statistics/Malta_Blue_Books/Pages/Malta-Blue-Books.aspx (accessed on 15 May 2016).

Royal Engineers Office, Malta (January 1910) Geographical Section General Staff [GSGS] No. 2430. *Map of the Island of Malta, Scale 1:31,680*. [Great Britain] War Office.

Royal Engineers Office, Malta (January 1910 with additions January 1923) Geographical Section General Staff [GSGS] No 2430. *Map of the Island of Malta, Scale 1:31,680*. [Great Britain] War Office.

Spiteri, S. (1996) *British Military Architecture in Malta*, Malta: The Author.

Thomson, D. (1966) *Europe since Napoleon*, London: Penguin Books.

Zammit, R. (2004) 'Japanese lieutenant's son visits war dead at Kalkara Cemetery', *Times of Malta*, 27 March, Valletta: Allied Newspapers Limited, 20–21.

Zarycki, T. (2007) 'An interdisiplinary model of centre-periphery relations: A theoretical proposition', *Studia Regionalne i Lokalne* (Special Issue): 110–130.

6 Asia's Great War

A shared experience

Xu Guoqi

This chapter addresses the kinds of experience that the peoples of China, Japan, India, and Vietnam shared in connection with the First World War. It moves beyond the national or even international level by presenting history from the transnational perspectives more truly characteristic of the era.[1] As I have argued elsewhere, "Shared experiences or past encounters are something different from a shared journey, which presumes a common destination and mutual interest despite possible difficulties, challenges, and tribulations along the way" (Xu 2014a: 1). Even given their many differences, the Chinese, Japanese, Indians, and Vietnamese did indeed have many experiences in common as they moved through the era of the Great War, not least through their direct involvement in the war.[2] All tried to use the Great War to promote their national development, international prestige or both. To various degrees, therefore, all four treated the war as an opportunity. The Allied countries turned to them for support and all made important contributions and the war and its immediate aftermath had a crucial impact in all four countries. They were treated as coloured peoples and for that reason received unequal treatment from the Western Powers. And they collectively experienced deep disillusionment with the results of the post-war conference and the resulting new world order. For all these reasons, the war and its aftermath marked a major turning point in each country's future development. It is also important to keep in mind that these Asian nations' involvement in the war clearly broadened and largely expanded the geographical and civilizational horizons of the Great War, and a shared history of Asia and the Great War therefore will bring the Asian voices and stories as well as Asian cultural geography to the otherwise possibly west-centric approaches of Great War studies.

Shared war experiences

As independent nations, China and Japan chose to get involved in the Great War. Both recognized an opportunity the European conflict presented and pursued it to advance their long-term national interests and international ambitions. In Japan's case, that interest was its ambition to subjugate China, while China's goal to enter the war, and especially take a seat at the post-war peace conference, grew from its determination to keep Japan out. Clearly a full understanding of either Chinese or Japanese participation requires bringing the other into account (Xu 2015). When war broke out in Europe in 1914, Chinese and Japanese were eager to step in as a way to redefine their positions in the world and reshape their national identities. The China-Japan connection in the Great War is one of tragedy, irony, and contradiction. These archenemies ended up working on behalf of the same side. They were rivals who spurred each other on and learned from each other, and each saw dealing with the other as the primary motivation behind their war policies and post-war agendas. Their investments in the Great War created forces for transformative change in the national development of both.

Japan had long been determined to become a powerful player in international politics and China's master. When the European War broke out in 1914 many influential Japanese saw it as the perfect opportunity to achieve both objectives. This was why Japan immediately jumped into the war that August and by November had taken Germany's Qingdao concession in Shandong. On January 18, 1915, Japan presented the Chinese government with the Twenty-One Demands. This move clearly revealed Japan's intention to turn China into a vassal state.[3]

Interestingly, like the Japanese, the Chinese also saw the outbreak of the Great War as an opportunity (Xu 2005). China's social transformation, as well as its more recent cultural and political revolutions, coincided with a war that provided the momentum and opportunity for China to redefine its relations with the world by inserting itself into the war effort. The war signaled the collapse of the existing international system and presaged a new world order, a development that fed China's desire to change its international status. But China also had immediate problems to deal with: its key reason for joining the war was to counter the now obvious Japanese aggression. Japan's presentation of the Twenty-One Demands in early 1915 only spurred Chinese determination to get involved in the Great War. Not surprisingly, Japan strongly opposed China's formal entry into the conflict, but the Chinese determination to link its fate with the Allied cause was so strong that they came up with a creative strategy, known as the "labourers-as-soldiers" scheme, as early as 1915 (Xu 2011). The scheme coincided with French and British search for new labour forces after heavy casualties they sustained in 1915 and 1916, respectively. Starting from 1915, China provided about 140,000 workers to the Western Front to support British and French war efforts (Xu 2011).

As with the other Asian peoples drawn into the war, Vietnamese experiences in the First World War marked a turning point in their history and the outbreak of the Great War provided Indochina momentum for change and transformation. The positions of French Indochina [Vietnam] and British India as colonies of European Powers differed from those of China and Japan. In the Indochinese case, the outbreak of war in Europe did not command much attention, and discussions and deliberations regarding the impact of the war on the country were limited and largely inconsequential. But many Indochinese, like the Chinese, had become deeply affected by the ideas of Social Darwinism at the turn of the twentieth century, and these spurred them to seek a new direction for their country by seeking national autonomy and national self-determination. If China and Japan were destined to get involved the Great War as early as 1895 in the wake of first Sino-Japanese war the Vietnamese were dragged into the orbit of European hostilities even earlier – in 1885, when Indochina became a French colony. When the Great War broke out, France turned to such colonies for support; the French government rallied to mobilize both human and material sources in Indochina, which would contribute enormously to the French war effort. Despite initial difficulties, Indochina increasingly contributed both human and material resources to the French military. According to Albert Sarraut (1923), the governor-general of Indochina during 1911–1914 and 1916–1919, military recruitment there started in 1915 with 3,000 men; in 1916, 36,000 Indochinese went to France; in 1917, the number was 9,922, for a total of 48,922 Vietnamese joining the military for France. These soldiers served in several different combat units, two were stationed in France, two in Macedonia, and one in Djibouti (Vu-Hill 2011a: 44). French recruitment of Indochinese labourers also started in 1915 with 4,631 men. In 1916, 26,098 labourers went to France; in 1917, 11,719 Vietnamese labourers arrived in France; in 1918 with 5,806 Indochinese arrived in France. In 1919, a further 727 Indochinese were brought to France for the post-war reconstruction effort, meaning that a total of 48,981 Indochinese labourers served in France during the war (Sarraut 1923: 42). There is also evidence that Vietnamese women went to France to serve there. A few volunteered to work in health services or at factories "by the side of our French sisters." There were even reports of female Indochinese in the workers' camps (Vu-Hill 2011b: 55).

India's involvement in the war was largely a by-product of its inclusion in the British Empire, and so not a decision made with India's own interest in mind. Its colonial master at first did not imagine it would need Indian help. The fighting, after all, was primarily between European peoples. But the British soon realized that they would have to mobilize Indian resources if they hoped to survive the conflict. About 1.2 million Indian men arrived in France during the First World War either as soldiers or labourers to work under their colonial masters near the imperial motherland. Their war experiences helped shape Indian understandings of their country's national fate and gave them new perspectives on the British and on Western civilization. To put it another way, Indian involvement, even under British direction, was important to Indians both in terms of national development and external relations. The war opened many Indians' eyes to the outside world and allowed them to dream and set high expectations as their so-called mother country fought a major war. They were witness to a world politics that was being transformed.

Although Indians and Vietnamese as subjects of Britain and France did not control their own war policies or even their own fates, and thus were not free to discuss, debate, and create independent policies around the war the way the Chinese and Japanese did, their fates were nonetheless deeply affected by the war and the post-war world order. Many Indian elites chose to support the British war effort with the longer view that it would help them achieve their dream of eventual independence. Indian involvement in the war under the British was important for at least two significant reasons. In the past, the world at large had meant little to most Indians; they had rarely given much thought to international and military affairs. Their involvement in the war for the first time confronted them with a fuller realization of their relation to the rest of the British Empire and the rest of the world. Second, India's part in the war had a deep impact on its elites' thinking about their country and on the rise of Indian nationalism. The Great War coincided with a major national awakening among Indians. Both Indian elite society and the lives of the lower classes would be affected in its course, making the war era a point of no return in India's national development.

Like the cases of China and Japan, Indians treated the war as an opportunity for national development. Santanu Das has suggested that Indian political opinion was unanimous in its support for the war. Though the war was universally seen as a catastrophe for Europe, it was seen as "India's Opportunity." This closely parallels the attitudes of the Chinese and Japanese. As P. S. Sivaswami, a member of the Indian executive council wrote,

> [The Indian's] loyalty is not the merely instinctive loyalty of the Briton at home or the Colonial, but the outcome of gratitude for benefits conferred and of the conviction that the progress of India is indissolubly bound up with the integrity and solidarity of the British Empire.
>
> (Das 2010: 351–352)

Likewise, in Vietnam, the most important legacy of the war was its effect on colonial Vietnamese society: it changed the mind-set of the Vietnamese, awakened their sense of nationalism and fed their personal growth with experiences of Europe and contact with men from other civilizations. The war in Europe certainly damaged French reputation in Indochina, as Albert Sarraut (1923: 37) wrote,

> The dreadful holocaust of our dead, the still-dark image of our richest provinces devastated, the procession of pains from which we will suffer for a long time to come, all these sorrows, in moments where our effort of will tries to divert the oppressive anguish, underlining, through the effect of contrast, one of the consoling consequences of the ordeal that we have endured: the Great War had the clear advantage of revealing the colonies to the French public.

According to Philippe M. F. Peycam (2012: 3),

> In the early twentieth century, Western pretensions of unchallengeable supremacy suffered a number of palpable blows: the Japanese victory over Russia in 1905, the butchery of World War I among European nations, the 1917 Soviet Revolution in Russia, and, in the Asian French colony, the persistence of opposition to colonial rule at both popular and elite levels.

For several years, the French did not do well in the war and this drove them to something close to concessions.

When the fate of France itself hung in the balance in Europe and when a less assured colonial state was trying to hold on to its position by conceding to the Vietnamese population limited, shared responsibility. The purpose was to obtain native support for the war effort and, beyond it, for the supposedly mutually beneficial project offered by French republican colonialism.

(Peycam 2012: 4)

This latter phenomenon arose against the background of major sociocultural transformations, most acutely experienced in the main urban centres of Saigon and Hanoi. Opposition crystallized into a historical moment in the midst of the First World War, "Arising from the aspirations of urban, Western-educated Vietnamese, this pursuit adopted original forms of activism, using newspapers as a distinct political force that flourished within the constraints of the colonial legal framework" (Peycam 2012: 6). As France and its empire became engulfed in the war, the colonial port city of Saigon found itself developing into a "space of possibilities". "Within its boundaries, a complex process of imposed acculturation and social interactions led to new expressions of Vietnamese consciousness on both an individual and a collective level" (Peycam 2012: 13).

Despite all the problems, challenges, and racism, living and working in Europe side by side with the French and other peoples provided the Vietnamese an opportunity to observe, to learn, and to understand different cultures and civilizations. Some had opportunities to interact with American soldiers when they came to each other's assistance during battles. The personal letters of Vietnamese soldiers reveal their impressions. For many Vietnamese soldiers the journey to France meant more than answering the call of their colonial master, it was also an eye-opening learning experience as they had opportunity to observe and interact with Westerners in their European homeland. It allowed them to compare and contrast the French with others and rethink their own national identity and position in the world. The future Vietnamese leader Ho Chi Minh went to Paris in 1919 to lobby for his country's independence and what happened to him there had significant impact on the Vietnamese search for a national identity and the future development of Indochina.

Of all the Vietnamese who were in France during the war, none was more important in transforming Indochina than Ho Chi Minh. Nationalists such as Ho, an almost-unknown person at the time, saw potential advantage in the war. When the fighting first broke out, he wrote to one of his friends and mentors,

Gunfire rings through the air and corpses cover the ground. The five Great Powers are engaged in battle. Nine countries are at war . . . I think that in the next three or four months the destiny of Asia will change dramatically. Too bad for those who are fighting and struggling. We just have to remain calm.

(Brocheux 2007: 12)

He concluded that he should go to France immediately to take the pulse of the larger world and better understand the role his country might play. He may have anticipated that the conflict would lead to the eventual collapse of the French colonial system (Duiker 2000: 51). Even though at the time Ho was engaged in undistinguished work, he worried about the fate of his country. In a poem composed in 1914, he wrote,

> In confronting the skies and the waters,
> Under the impulse of will that makes a hero
> One must fight for one's compatriots.

(Cited in Duiker 2000: 53)

Due to their shared view of the Great War as an opportunity, these four nations all found ways to make significant contributions to the war effort. Apart from recounting the actual fighting that took place in China – notably the clashes that occurred in Chinese territories when the German forces stationed there were attacked by the British and the Japanese – historians must recognize the sacrifice of the hundreds of thousands of Asian men who served (and died) on the Western Front. Japan not only helped expel Germany from Asia, it provided crucial navy support for the Allied side. China's involvement was perhaps the most unusual. The program to send Chinese labourers to the Western Front marked the first time in modern history that the Chinese government took the initiative in affairs distant from Chinese shores. Since the Great War was a 'total war', fought as a terrible trench war on the battlefield and on the home front where all hands helped support the fighting, it consumed massive numbers of fighting men and other resources. The enormous human resources contributed by the Chinese, Indians, and Indochinese must be counted as an important part of this global war effort. China's 140,000 workers clearly helped the British and French maintain their fighting strength and provided crucial logistical support. Both India and Indochina provided significant numbers of soldiers and labourers in support of the British and French. Thousands of men from these Asian countries died in the conflict.

In addition to the substantial human resources contributed by Asians, they provided other important assistance to the Allied side as well. For instance, Indochina made other sacrifices for the French war effort. By the end of 1916 alone, it had loaned France more than 60 million francs, alongside goods valued at approximately 30 million francs. Another figure puts the French government's wartime loans from its colony at over 167 million francs between 1915 and 1920, and the sale of the war bonds in Indochina reached 13,816,117 francs. Indochina lent France more than 367 million of the 600 million francs borrowed from all its colonies in Asia and Africa. Significantly, 30 per cent of that amount came from native Vietnamese individuals. According to Albert Sarraut (1923), Indochina's total cash contribution between 1915 and 1920 amounted to 382,150,437 francs; donations by individuals amounted to 14,835,803 francs. The Vietnamese sent cash donations of 10 million francs to aid war victims and contributed 11,477,346 francs to the 1915 military campaign (Vu-Hill 2011a: 37–38). War aid also included raw materials such as coal, rubber, and minerals; dry goods such as rice, tea, tobacco; fabrics such as cotton and khaki; and clothing. Over the course of the war, Indochina supplied France with 335,882 tonnes of rice, corn, alcohol, beans, tobacco, cotton, rubber, copra, timber, cooking oil, and lard. As with colonial India, Indochina had to ensure the wages, pensions, and family benefits for the men who had been mobilized, as well as pay the expenses of the Indochinese military hospital. The latter added up to 4,040,000 francs sent to France in 1917 and 1918 (Brocheux 2007: 19). Once the war was over, Indochina also helped to pay for the reconstruction of five cities that had been destroyed – Carency, Origny-en-Thierache, Hauvigne, Chauvignon, and Laffaux.

If Asian participation had made the war truly great and worldwide in scope, their witnessing the European carnage and struggle had other, unexpected effects. The Chinese, Indian, and Vietnamese experiences, through their encounters with Westerners in Europe, collectively broadened their perspective and generated new thinking about Eastern and Western civilizations. Indian and Vietnamese participation in the war would contribute to their workers' and soldiers' political awakening, new national thinking, and confidence. India's eventual independence and democracy drew on the new confidence that they could stand as equals to their former colonial master, while fallout from the aftermath of the war planted the seeds of China's and Indochina's eventual conversion to communism. The shared excitement among the war's Asian participants at the prospects of the post-war peace conference also indicates the war's importance for Asia.

The idea of a shared history across Asia in this period makes perfect sense: The war itself, later called a "world war," brought together peoples who would have otherwise had little encounter. And not only did elites share interest in the war and ideas of national development, in the sense that they gained inspiration locally and internationally, but even the most marginalized of their countrymen – the workers and foot soldiers recruited from among the poor in China, India and Indochina – personally experienced life in foreign places and cultures, often under dire circumstances. Asian involvement in the First World War is thus a unique chapter in both Asian and world history, as Asian participation transformed the meaning and implications of the broader conflict.

The First World War transformed domestic politics of each nation and the circumstances in each of these countries had changed dramatically during the war years. China struggled to come together as a nation, while India started its long journey to independence. China and Indochina would eventually follow a socialist path, but in Japan, the war gave rise to a new sense of national pride that eventually led the Japanese to adopt military methods and challenge the West outright. Asian involvement in the war gave the Allied side both moral prestige and a strategic and human resources advantage. But the fallout from the First World War across Asia was tragedy, paradox, and contradiction. The Great War was about imperial ambitions, but in the same era, China destroyed its own empire in its desire to become a modern republic and a nation-state. Japan used the war to strengthen its claim to empire status, while the Indian and Indochinese experiences inspired working toward the casting out of their imperial masters in order to become independent nations.

Asians' shared disappointments at the Paris Peace Conference

The four countries under consideration here shared soaring expectations for the war and humiliating disappointments in its aftermath. I argue that those expectations, frustrations, and the disappointments of the post-war peace conference were a striking collective experience, although few scholars have attempted to compare them directly. All four were enthusiastic about the new world order laid out in Woodrow Wilson's "Fourteen Points" speech.[4] The relatively weak Chinese, Indians, and Indochinese hoped to gain an equal voice at the post-war peace conference to present the case for their national aspirations. The Japanese expected that the Paris Peace Conference would seal their long-cherished aspiration to be recognized as the dominant Asian power and give their recently gained interests in China an international stamp of approval. More importantly, the Japanese hoped that Western Powers would finally accept Japan as a full equal.

For colonial India and Indochina, Wilson's ideas of national self-determination generated great excitement. In 1919, Ho Chi Minh, then known as Nguyen Ai Quoc (literally "Nguyen who loves his country"), first made his presence known in Paris. This largely unknown Indochinese colonial became very active in Paris. He believed that this was the time for his people to pursue definitive autonomy and perhaps eventual independence. In cooperation with other countrymen in Paris, Ho prepared a petition for Indochinese autonomy that was presented to the conference. That petition was presented by a "Group of Annamite Patriots in France," though the founding of this group is undocumented (Ho Tai 1992: 69). The petition was clearly influenced and motivated by Wilson's ideas, but it was not at all politically radical. It did not ask for independence but rather for autonomy, equal rights, and political freedoms. It called for the following:

1. General amnesty for all native political prisoners;
2. Reform of Indochinese justice by granting the natives the same judicial guarantees as Europeans;
3. Freedom of the press and opinion;
4. Freedom of association;
5. Freedom of emigration and foreign travel;
6. Freedom of instruction and the creation in all provinces of technical and professional schools for indigenous people;
7. Replacement of rule by decree with the rule of law; and
8. Election of a permanent Indochinese delegation to the French parliament, to keep it informed of the wishes of indigenous people.

But the Vietnamese petition to the conference principals did not succeed and in fact was ignored. It was considered "too obscure" to receive an answer from Wilson or the French principal Clemenceau. Although his petition was ignored by the powers, Ho was taken quite seriously by fellow Asians and the world started to pay attention to him during and subsequently after the Paris Peace Conference.

His experiences in Paris turned Ho to communism and eventually made Vietnam a socialist country in much the same way that China had turned to communism after its disillusionment with the West. Ho Chi Minh would later claim he had concluded, "The liberation of the proletariat is the necessary condition for national liberation. Both these liberations can only come from communism and world revolution" (Ho 2007: 5–6). Ho found in the ideology of Lenin and the communist international the "path to our liberation" (Hess 1990: 15). In one of his articles on "Indochina," he wrote, "The tyranny of capitalism has prepared the ground: the only thing for socialism to do is to sow the seeds emancipation" (Ho 2007: 3–4). Ho officially joined the Communist Party in 1921, though his turn to communism was more motivated by desire to achieve national independence rather than for ideological objectives.

As Ho sought a path to champion Indochinese nationalism in an international setting, large numbers of Vietnamese soldiers and workers who had served in the war were charting their own courses. Those who returned home began organizing, leading, and agitating within the 1920s labour movement. Ton Duct Hang, a sailor in the French fleet during the Great War who would later become president of the Democratic Republic of Viet Nam, used his knowledge of organized labour to found the first Association of Workers of the Saigon Arsenal in 1920. Nguyen Trong Nghi, a worker in a French factory during the war, became an agitator and leader in the workers' demonstrations in Nam Dinh in the late 1920s. "The 1921 reform of the village political system in Tonkin and the emergence of a modern organized labour movement in the late 1920s were direct consequences of the war on French colonialism in Indochina" (Vu-Hill 2011a: 10–12). Those who remained in France formed the earliest "Indochinese colonies in France" (Vu-Hill 2011a). These communities grew in cities throughout France in the 1920s, and the diaspora there formed a significant political block. In the decades between the two world wars, a number of these men engaged in political struggles to liberate Indochina, and as Kimloan Vu-Hill (2011a: 9) has suggested, "without World War I, the path to revolution and to national independence would have taken a different direction". World War I indeed marked a turning point in the history of French Indochina and Vietnamese national development.

As mentioned earlier, due to their link of war efforts to Indian national fate, nearly all prominent Indian politicians supported the British war effort. Mohandas K. Gandhi was one of them. But Gandhi soon became one of India's towering figures in its cause of national independence. During the wartime, Gandhi cautioned that "it was more becoming and far-sighted not to press our demands while the war lasted" (Das 2011: 73). Given India's significant contributions to war and the British Empire and its high expectations for the post-war world order, it was natural that for Gandhi and other Indian nationalists looked forward to their representation at the peace conference. When the Indian National Congress convened in December 1918 for its annual session, "in view of the pronouncements of President Wilson, Mr. Lloyd George, and other British statesmen, that to ensure the future peace of the world, the principle of Self-Determination should be applied to all progressive nations," it adopted a resolution that called for the application of those principles to India and demanded that India be recognized by the powers as "one of the progressive nations to whom the principle of self-determination should be applied" (Manela 2007: 96). The Congress further urged that elected delegates represent India at the peace conference. Other organizations involved in the home-rule movement, such as Annie Besant's All-India Home Rule League congratulated the British sovereign on the Allied victory but demanded as "absolutely essential" the immediate implementation of home rule in India. India, of course, was bound to be disappointed since Britain was unlikely to give up India or any of its colonial possessions.

But the demands of Indian nationalists went nowhere. Britain reneged on its promise of self-rule and resumed repressive policies after the war. There was little surprise that this met with a sense of disappointment amongst much of the Indian population and beyond. Gandhi became more and more disillusioned with British government after 1919. He concluded that the British government "is immoral, unjust and arrogant beyond description. It defends one lie with other lies. It does most things under the threat of force. If the people tolerate all these things and do nothing, they will never progress" (Masselos 1991: 163). Gandhi thus "shifted in 1919 from a position of firm if critical support for Indian membership in the British Empire to one of determined opposition to it" (Manela 2007: 9). Winston Churchill considered Gandhi a "seditious fakir" (Masselos 1991: 151). The historian A. Rumbold (1979) called the period between the outbreak of the Great War and 1922 when Gandhi's campaign of non-cooperation came to an end with his arrest a watershed in the history of British power in India. The future Chinese communist leader, Mao Zedong, observed in 1919 that at the Paris Peace Conference India had "earned itself a clown wearing a flaming red turban as representative" and "the demands of the Indian people have not been granted" despite the fact that India had risked "its own life to help Britain" during the war (Schram 1992: 335). The Indian nationalist movement went through a sea change thanks to the war. The Indian National Congress had been a pillar of the empire until 1914, but once the war was over, it became a determined enemy. One can argue that India's experience in the Great War and at the peace conference set it on the path toward full independence after the Second World War.

In Paris, Japan won what it wanted most – Great Power status and the province of Shandong – but even so the Japanese delegates came away disappointed. Many in Japan were critical of what was acknowledged by those within the country to be a failure. Baron Saionji apologized in his formal report to the emperor: "I am sad that we could not accomplish our wishes in total" (Schlichtmann 2009: 124). Japan's disappointments and disillusionments with the peace conference can be split into several categories. Japan's biggest disappointment was the dismissal of its racial equality proposal; meaning that the Western Powers' refused to accept the Japanese as true equals in the Western-dominated world order. Japan would remain outside the white power club and would continue to share second-class status with fellow Asians; such disillusionment helped to give rise to Japan's later go-alone policy and expansionist drive into China.

Japan experienced other disappointments that stemmed from the war and the peace conference. First of all, Japan was going through something of a national identity crisis. While the Chinese found the nineteenth-century world order terribly wrong, unfair, and hostile, Japan had seen herself as the "pioneer of progress in the Orient" for successfully adopting the material advances of Western civilization, with special determination to emulate Germany. After the Meiji restoration, Japan adopted practices after the new German political system, the German constitution, and the military training system. The new Japanese national regime tried to restructure public life and institutions after the Prussian model, including education, medicine, and science, in addition to the political and military institutions. In Germany, Japan had been known as the "Prussia of the East" after its defeat of China in 1895. As Bernd Martin points out, "The option was clear; in the young and aspiring German Empire the Japanese saw the model of an orderly and politically stable society, or patriotism and, of course, the people's loyalty towards their monarch" (Martin 1995: 35). The German defeat in the Great War and the imposition of a new world order forced Japan to conclude that it might have followed the wrong model – after all, Germany was now a denounced nation. Germany's defeat, the collapse of its monarchy, and its adoption of a republican constitution had fundamentally shaken Japan's faith in constitutional monarchy and led to a sort of national identity crisis. Although the Japanese military largely stuck to the German model, Germany's defeat nonetheless caused many in Japan to rethink the political implications of this model.

The Great War also markedly increased the Japanese sense of diplomatic isolation. Prince Konoe Fumimaro perhaps expressed the deep disappointment after the war best. Konoe, who himself attended the peace conference and came away unhappy with the new world order and Japan's place in it, saw how the Western Powers commandeered the entire course of the peace conference (Oka 1983: 14–15). When he returned to Japan, he found it a sad country. "Everything I see and hear makes me unhappy" (Oka 1983: 17). The Japanese delegates came away from Paris convinced that the United States was out to stop them in China and that the world did not trust Japan. According to Harvard historian Akira Iriye (1997: 50),

> The main challenge facing Japanese diplomacy after the World War was how best to define its ideological foundations now that the old diplomacy of imperialism was giving way to novel approaches being promoted by the United States, Russia, China and other countries.

After the war, American policy aimed to re-establish order and stability in East Asia, which mandated that Japan return Shandong to China. America also wanted to demolish the existing system of imperialist diplomacy in East Asia, which seemed specifically targeted at the Japanese (Iriye 1990: 13–14). The Anglo-Japanese alliance, which had been the cornerstone of Japanese diplomacy since 1902, was also dismantled. Japanese foreign minister Uchida Yasuya was extremely upset when he learned about the British intent to terminate it (Gates

2014: 75). According to Tadashi Nakatani (2014: 171–172) at the peace conference, Japan insisted on the old diplomacy and "in taking this position, the peace meant only the end of the 'one chance in a thousand' to Japan".

Even Japan's initial victory over its possession of Shandong proved costly. Like Japan, China was obsessed with Shandong, and while Japan did its best to keep it, China was determined to get it back. At the news of the Powers' refusal to support the return of Shandong to China, Chinese students took to the streets on May 4, 1919, in what eventually became known as the May Fourth Movement, a turning point in modern Chinese history. The Chinese felt betrayed and became deeply disillusioned with the liberal West. "The betrayal at Versailles" led many Chinese elites to doubt the value, and even the possibility, of China's identifying with the West. Yan Fu, a famous scholar and translator of many Western books, declared that the behavior of the West in 1919 showed that "three hundred years of evolutionary progress have all come down to nothing but four words: selfishness, slaughter, shamelessness, and corruption" (Pusey 1983: 439). In his 1920 annual report to Earl Curzon, the British minister to China, Alston wrote,

> The rising tide of international esteem began to flow when China refused, weak as she was, to be bullied into signing the Treaty of Versailles. Though the momentary political victory at that time went to Japan, the moral victory remained with China, and has since culminated in her obtaining one of the temporary seats on the Council of the League of Nations.[5]

The May Fourth Movement marked the end of all-out efforts to join the liberal Western system began by China seeking to join the First World War. This break prompted a search for a third way, a way between Western ideas and Chinese traditional culture. May Fourth was an expression of the public's disillusion with the Western Powers' new foreign policy, after their refusal to return Shandong. "Young China's faith in Wilsonian idealism has been shattered to dust. 'The New World Order' is no more," wrote one Chinese author.[6] For Chen Duxiu, a pro-West liberal scholar who later a co-founded the Chinese Communist Party, Wilson had turned out to be an "empty cannon" whose principles were "not worth one penny."[7] Chen (1965: 629) wrote, "It is still a bandits' and robbers' world, it is still a world where justice is overpowered by might". Students across China openly expressed their disdain at the failure of Wilsonianism. Those at Beijing University cynically joked that Wilson had discovered a jolting new formula for the idealistic world order: "14 [referring to Wilson's Fourteen Principles] = 0" (Zhong 1979: 222).

The humiliation China suffered in Paris put a damper on the pursuit of a Westernized national identity. The practical failure of the Paris Peace Conference alienated Chinese intellectuals, many of whom had been exposed to ideas about the decline of West generated by the likes of Oswald Spengler. After their heartbreaking experience at the peace conference, Chinese began to view the Bolshevik Revolution as the only successful model for state building, and Bolshevik Russia was the only power that appeared sympathetic to Chinese aspirations. Mao

Zedong, then just another young educated youth, concluded that Russia was "the number one civilized country in the world" (Fairbank and Feuerwerker 1986: 802). And, as Benjamin Schwartz (1968: 286) explains,

> Paradoxically, one can actually assert that one of the main appeals of Marxism-Leninism to young Chinese was its appeal to nationalistic resentments. The Leninist theory of nationalism provided a plausible explanation for China's failure to achieve its rightful place in the world of nations.

Chinese disappointments in Paris ran much deeper since China had pinned so many hopes on the post-war world order. The Chinese had carefully prepared for the peace conference since 1915 because they knew their weak and disrespected country had little other leverage with the Great Powers. With its official declaration of war against Germany and the large contingent of Chinese labourers sent to Europe to support the Allies, China had earned its place at the conference, but only as a third-rank nation having two seats. Japan claimed five and in retrospect, any Japanese success at the conference automatically meant failure for China. Even so, the Chinese capitalized on the opportunity and managed to inject substantially new perspectives into conference discussions. But neither did they realize their dream of equality with other nations nor the recovery of Shandong. By turning to socialism, China and Indochina would go their own way.

Conclusion: the World War as a collective turning point

Asia in the year 1919, in the wake of the European War, was fundamentally different from the Asia of 1914 – socially, economically, intellectually, culturally, and ideologically. The broadly defined years of the Great War coincided with a period of tremendous change as the old Confucian civilizations began to collapse. China and Japan struggled toward modern nationhood and sought to assume equal relations with the West. India set out on its long journey toward independence while China pursued internationalization and national renewal. While China and Indochina would turn to communism, the Great War led to the emergence of a Japan that had become prosperous and strong enough to eventually use military force to challenge the exclusionary West.

The war produced sometime bizarre defeats and victories. China had joined the side of victors but saw few rewards. Japan was a victor whose status in the world improved substantially, but its gains lit the fuse for its future destruction. The Great War brought about an end to the nineteenth-century world system that had kept Indochina and India colonial backwaters, presenting an opportunity for a general re-ordering of world affairs. Educated Asians understood that the war represented the moral decline of Europe, but the post-war world system could not dislodge the entrenched Powers and would deliver little beyond collective disappointment in the war's immediate aftermath.

Asian involvement made the Great War into the First World War, and the First World War changed the world. The Asians' part in the war and the part the war played in the collective development of Asia represent the first steps toward full national independence and international recognition among the various peoples across the region. The war had a powerful impact on national identity: for Japan, it provided confirmation as a world power, for China, the war finally sparked a cultural and political revolution and kindled the desire for both a stable republic and international recognition. The war clearly sparked visions of independence in India and Vietnam.

The Great War, with the radical opportunities it provided for people from distant parts to be together and interact, also introduced the notion of fusing civilizations, East and West. It shaped the thinking and political ideas of future Asian leaders – Mao Zedong, Ho Chi Minh, Gandhi, and Japan's Konoe Fumimaro. These leaders all would play an active political role or be somehow involved in the Second World War and the new world order that resulted from that conflict.

The story of Asia in the First World War is full of irony, paradox, and contradictions. Indians enthusiastically supported the war efforts of their imperial rulers, but with an eye to realizing their dream of independence. The issue of neutrality played out in contradictory ways: Britain, used the German violation of Belgian neutrality to legitimize its entry into the war, but it also intentionally violated the neutrality of China and supported Japanese aggression there. The war had everything to do with empires, but while China dissolved its own in its struggle to become a republic and a nation-state, it faced repeated attempts to revive the monarchy, first in the person of the president Yuan Shikai and then by the former Manchu emperor.[8] Neither lasted long. The war promoted Japan to the top ranks of world power, but its aftermath only increased the Japanese sense of mistreatment and frustration. Japan was badly stung by the failure of its racial equality initiative at the peace conference and the worldwide denunciation of its empire building aspirations.

The war affected Japan's domestic politics and international relations and played an important role in shaping Japan into an imperial aggressor. We can see that no foreign policy initiative had a stronger impact on China's domestic politics than that on its entry into the First World War. But instead of enjoying the fruits of its first major diplomatic initiative, the Chinese people tasted only bitter social disorder, political chaos, and national disintegration. Disputes around war participation fed the flames of factionalism, encouraged warlordism, and put China on the road to civil war. For India and Indochina, the war's political repercussions helped inspire their struggle toward independence and modern nationalism. Indians and Indochinese took the occasion of the war, and the discredit it brought on Europe, to work toward escaping their imperial masters' control. All across Asia, people saw the war and the post-war peace conference as a moment in which they could advance their national development and international status, and all had high hopes for the post-war world. All of them ended up deeply disillusioned but equally determined to find better weapons to advance their cause.

Notes

1 This chapter draws on materials from my recent book 'Asia and the Great War: A Shared Perspective' (Oxford, UK: Oxford University Press, 2017). For recent comprehensive

studies of the war from a transnational perspective, see Winter (2014) and Gerwarth & Manela (2014).

2 Of course, many other Asian nations such as Singapore, Thailand (Siam), and Malaysia were affected by and even got involved in the war one way or another. Turkey geographically is an Asian nation and was critically important in the Great War, but its story deserves its own volume. Due to space and thematic constraints, this chapter chose to focus on four most representative ones: India (British colony), Vietnam (French colony), China (a troubled but ambitious nation) and Japan (a rising power which tried to fish in the troubled water).

3 For details, see Xu (2005: 93–97). For studies on how the Japanese responded to the war and war's impact on Japan, see Dickinson (1999, 2013, 2014).

4 For details on Wilsonian ideals and their impact, see Manela (2007).

5 British National Archives: FO405/229/2.

6 Hu (1919: 346–7), cited in Zhang (1991: 74).

7 *Meizhou Ping lun,* no.20 (May 4, 1919).

8 For details on this, see Xu (2014b).

Bibliography

Brocheux, P. (2007) *Ho Chi Minh: A Biography*, translated by Chaire Duiker, New York: Cambridge University Press.

Chen, D. (1965) *Duxiu Wencun* (Surviving Writings of Chen Duxiu), Hong Kong: Xianggang Yuandong Tushu Gongsi.

Das, S. (2010) 'Ardour and anxiety: Politics and literature in the Indian Home Front', in Heike Liebau, Katrin Bromber, Katharina Lange, Dyala Hamzah, Ravi Ahuja (eds.), *The World in World Wars: Experiences, Perceptions and Perspectives from Africa and Asia*, Leiden: Brill, 351–352.

Das, S. (2011) 'Indians at home, Mesopotamia and France', in S. Das (ed.), *Race, Empire and First World War Writing*, Cambridge, UK: Cambridge University Press, 70–89.

Dickinson, F.R. (1999) *War and National Reinvention: Japan in the Great War, 1914–1919*, Cambridge, MA: Harvard University Press.

Dickinson, F.R. (2013) *World War I and the Triumph of a New Japan, 1919–1930*, Cambridge: Cambridge University Press.

Dickinson, F.R. (2014) 'Toward a global perspective of the Great War: Japan and the foundations of a twentieth-century world', *American Historical Review*, 119(4): 1154–1183.

Duiker, W. (2000) *Ho Chi Minh*, New York: Hyperion.

Fairbank, J.K. and Feuerwerker, A. (eds.) (1986) *The Cambridge History of China: 1912–1949*, 13, part 2, Cambridge, UK: Cambridge University Press.

Gates, R. (2014) 'Out with the new and in with the old', in T. Minohara, T.-K. Hon and E. Dawley (eds.), *The Decade of the Great War: Japan and the Wider World in the 1910s*, Leiden: Brill, 64–82.

Gerwarth, R. and Manela, E. (eds.) (2014) *Empires at War*, Oxford: Oxford University Press.

Hess, G.R. (1990) *Vietnam and the United States*, Boston: Twayne Publishers.

Ho, C.M. (2007) *Down With Colonialism*, New York: Verso.

Ho Tai, H.-T. (1992) *Radicalism and the Origins of the Vietnamese Revolution*, Cambridge, MA: Harvard University Press.

Hu, S. (1919) 'Intellectual China 1919', *Chinese Social and Political Science Review*, 4(4): 346–347.

Iriye, A. (1990) *After Imperialism: The Search for a New Order in the Far East, 1921–1931*, Chicago: Imprint Publications.

Iriye, A. (1997) *Japan and the Wider World: From the Mid-Nineteenth Century to the Present*, London: Longman.

Manela, E. (2007) *The Wilsonian Moment: Self-Determination and the International Origins of Anticolonial Nationalism*, New York: Oxford University Press.

Martin, B. (1995) *Japan and Germany in the Modern World*, Providence: Berghahn.

Masselos, J. (1991) *Indian Nationalism: A History*, New York: Sterling.

Nakatani, T. (2014) 'What peace meant to Japan: The changeover at Paris in 1919', in T. Minohara, T.-K. Hon and E. Dawley (eds.), *The Decade of the Great War: Japan and the Wider World in the 1910s*, Leiden: Brill, 168–188.

Oka, Y. (1983) *Konoe Fumimaro: A Political Biography*, translated by Shumpei Okamoto and Patricia Murray, Tokyo: University of Tokyo Press.

Peycam, P.M.F. (2012) *The Birth of Vietnamese Political Journalism: Saigon 1916–1930*, New York: Columbia University Press.

Pusey, J. (1983) *China and Charles Darwin*, Cambridge: Harvard University Press.

Rumbold, A. (1979) *Watershed in India, 1914–1922*, London: Athlone Press.

Sarraut, A. (1923) *La Mise en valeur des Colonies Françaises*, Paris: Payot.

Schlichtmann, K. (2009) *Japan in the World: Shidehara Kijuro, Pacifism, and the Abolition of War, Lanham*, MD: Lexington Books.

Schram, S.R. (ed.) (1992) *Mao's Road to Power: Revolutionary Writings, 1912–1949*, Armonk, NY: M.E. Sharpe.

Schwartz, B.I. (1968) 'Chinese perception of world order', in J.K. Fairbank (ed.), *The Chinese World Order: Traditional China's Foreign Relations*, Cambridge, MA: Harvard University Press, 276–288.

Vu-Hill, K. (2011a) *Coolies into Rebels: Impact of World War I on French Indochina*, Paris: Les Indes Savants.

Vu-Hill, K. (2011b) 'Sacrifice, sex, race: Vietnamese experiences in the First World War', in S. Das (ed.), *Race, Empire and First World War Writing*, Cambridge, UK: Cambridge University Press, 53–69.

Winter, J. (ed.) (2014) *The Cambridge History of the First World War*, Cambridge: Cambridge University Press.

Xu, G. (2005) *China and the Great War: China's Pursuit of a New National Identity and Internationalization*, Cambridge: Cambridge University Press.

Xu, G. (2011) *Strangers on the Western Front: Chinese Workers in the Great War*, Cambridge, MA: Harvard University Press.

Xu, G. (2014a) *Chinese and Americans: A Shared History*, Cambridge, MA: Harvard University Press.

Xu, G. (2014b) 'China and Empire', in R. Gerwarth and E. Manela (eds.), *Empires at War: 1911–1923*, Oxford: Oxford University Press, 214–234.

Xu, G. (2015) 'The Great War in China and Japan', in O. Frattolillo and A. Best (eds.), *Japan and the Great War*, Basingstoke: Palgrave Macmillan, 13–35.

Xu, G. (2017) *Asia and the Great War: A Shared History*, Oxford: Oxford University Press.

Zhang, Y. (1991) *China in the International System, 1918–20: The Middle Kingdom at the Periphery*, New York: St. Martin's Press.

Zhong, G. (1979) *Wu Si Dong Hui Yi Lu (Recollections of the May fourth Movement)*, Beijing: Zhong guo she hui ko xue chu ban she.

Part 2

Commemorative spaces

7 The art of war display[1]

The Imperial War Museum's First World War galleries, 2014

James Wallis and James Taylor

London's Imperial War Museum (IWM) remains an institution with few comparisons.[2] Its uniqueness lies in it having been established to commemorate a historical event yet to be resolved, and far from any kind of conclusion. Furthermore its genesis as a national museum was the result of a propaganda initiative, 'borne out of the sustaining ideology of war' (Jones 1996: 158). By the time that the 1918 Armistice had brought an official end to the fighting against Germany, the IWM had already taken to the task of establishing a national memory of this momentous event. It did this by recording the involvement of all levels of society drawn into its all-encompassing nature, thereby generating a resonance from the stories that its objects told. The museum's founders were, for that reason, undertaking something without precedent – achieved through a policy of delegated contemporary collecting of objects that would 'speak most eloquently of the human experience of war' (Cornish 2004: 49). What has since been termed a 'radically different approach to collecting and recording' (Kavanagh 1990: 20) lay in the institution's collective determination to make its resources relevant to the wartime population, with its efforts fittingly determined on the democratic acquisition of items 'redolent of the involvement of the common man and woman in the war, whether at the Front or at home' (Cornish 2004: 38).[3]

The museum opened at the Crystal Palace in South London in June 1920 amidst a strong-felt desire 'to consign the unprecedented "War to end all Wars" to history' and 'to look back upon War, its instruments and its organisations, as belonging to a dead past' (Milner 2002: 11). One commentator advocated the museum's early success as mirroring:

> Its public acceptance and emotional appeal as a site of memory and mourning, serving both the nation and the empire. The making of this institution signifie[d] the search for a symbolic form that could express collective bereavement and celebrate imperial unity at the same time.
>
> (Goebel 2007: 167)[4]

With the passing of time, Britain's War Museum continued to fulfil its primary purpose in providing opportunities for the public to study and understand the history of the First World War. The collections it held made it a storehouse, or a 'repository of memory'. Implicit within that notion is 'of memory objectified, not belonging to any one individual so much as to audiences, publics, collectives and nations, and represented via the museum collections' (Crane 2000: 2). In this way, the museum's continued existence, and the way in which it established its exhibitions, therefore operated as a facility for simultaneously preserving, shaping and constituting understanding of the First World War.

We now seek to position the institution within the context of more recent debates raised by works within the field of museum studies. Broadly speaking, this discipline has defined museums as multi-functional sites of both knowledge production and interaction between people and the material world.[5] Analysis of their societal function identified them as credible institutions with civic educational and socializing purposes, traditionally linked to the nation-state.[6] In response to feminist, postmodern and postcolonial critiques, museological collections and processes were radically re-signified – in a 'postmodern shift from master discourses to the horizontal, practice-related notions of memory, place and community' (Andermann and Arnold-de Simine 2012: 3). Keeping this understanding in mind, we turn to the multi-layered nature of exhibitions, in outlining ideas from existing literature, to account for their embedded politics.

The politics of display

Till (2001: 276) defines exhibitions as 'theatrical, staged spaces that perform selective versions of the past'. Curators can accordingly be termed authors who interpret the past by relating space, objects and written text in distinct combinations. These, in turn, are interpreted by visitors in various ways. The function of a historical exhibition stands as a medium of systematic yet mediated communication, an experiential space of 'material speech' (Ferguson 1996: 182; Luke 2002; Lisle 2006). Within such a context, museums construct and transmit meaning for visitors as 'intentional communicative acts' (Mason 2005: 204), through their interpretation of material evidence. They 'articulate or reinforce frameworks of knowledge – known as discursive formations – in display' to convey 'validity upon objects' (Newman and Mclean 2006: 57). Situated within the context of a historical museum, such institutions are instructed with defining a particular, constructed representation of past events for public consumption – so as to validate a nation's history (Trofanenko 2010: 270). In this vein, an exhibition can be considered as a system of representation that aims to convert audiences to particular values or intended messages.[7] Such a selection process is regulated and political in what is included or excluded, because meaning-making has 'the power to define, legitimize, enforce, negotiate, claim or oppose certain meanings' (Mason 2005: 208).[8] Exhibitions are duly able to confer legitimacy on specific interpretations of history, because strategies of display can both empower or dis-empower particular groups. There is a resulting academic interest in critiquing this controlled practice of exhibition-making: 'What are the processes, interest groups and negotiations involved in constructing an exhibition? What is ironed out and silenced? And how does the content and style of an exhibition inform public understandings?' (Macdonald 1998: 1, 10).

The impact of new museology during the late 1980s and early 1990s brought about significant changes in how state-engineered museums enacted their approach, message and practice. From this, museum environs were re-cast as spaces of community and public engagement, whilst exhibitions took on a more populist stance (Message 2006; Crooke 2007).[9] Narrative displays were required to provide 'social and learning experiences' with 'opportunities for visitors to engage in critical thinking and questioning ... to reach their own conclusions' (Kelly 2007: 287). Subsequently the historical museum professional's role was slowly reworked, so as to make public history '[as] shaped by scholarship and research, by the need to make that information accessible, and by an obligation to be of service to the public' (Yeingst and Bunch 1997: 155).

A historical museum enables visitors to experience the past from a personal perspective, derived from the notion that its displays can 'contribute significantly to the construction of personal and shared identities' (Watson 2007: 160). An institution of this ilk operates as a site of 'public memory' defined by 'the interplay between official, commercial and vernacular memory' (Hodgkin and Radstone 2006: 172). Within the context of the state-sponsored IWM, its curators remain 'under pressure to produce exhibitions that portray national history in a celebratory tone and produce a shared national identity that excludes controversy and difference, affirms civic pride and forms better citizens' (Cameron 2007: 337).[10] Such discursive constructions brought about tensions that curators and institutions had to address:

> How can objects be used to communicate history without presenting history didactically? How can museums return power to objects without making the experience merely aesthetic? Does the apparently unique position of history museums mean that they will remain didactic and objective, giving visitors the historical narrative through which to understand the objects, while other types of museums minimise interpretive text.
>
> (Lord 2007: 355)

Hereby an explanatory narrative of the First World War – as situated within the IWM – needed to remain universal in 'subsuming the experiences and histories of different communities into the one historical experience of the nation as a single community' (Witcomb 2003: 155). The institution remained obliged to perform an official view of the past, in 'producing narratives which form an integral part of national identity politics' (Andermann and Arnold-de Simine 2012: 9). A combination of these criteria would all play an influential role in fostering new displays at IWM London.

The 2014 First World War galleries

Having situated this case study within some key contextual debates, we now consider the First World War galleries that opened at IWM London in 2014. We apply an understanding of this exhibition environment as created 'through a complex of practices and systems of knowledge' (Macleod 2005: 1) – making use of a geographically informed lens to explore the malleability of this particular museum space.

Roppola has advocated that 'design is, by definition, intentional', meaning that museums are 'fabricated stages, performance spaces for pedagogical and political purposes' (Roppola 2012: 11–12). Producing galleries is, therefore, an act of negotiating – in terms of simultaneously suppressing and empowering – a constrained, finite space made readable for public consumption. Such acts can never be definitive, as already outlined – indeed respecting the institution's past practice meant that 'only certain approaches were to be possible [. . . which] defined the horizons of the thinkable from the very beginning' (Macdonald 2002: 214). The museum's decision makers had already expressed a strong desire for the institution to be centrally positioned for the national 2014–2018 First World War Centenary commemorations. By necessity, this set a natural yet high-profile deadline for the completion of new long-term galleries, by virtue of the anticipated media interest generated by these anniversaries.

As noted earlier, broader critique of museums had hardened the notion that public demand and expectation would be determining criteria within any new exhibitions (McManus 2011). Such a backdrop has witnessed museums operating with more of a consultancy-based mind-set, ensuring that their prospective product would take account of external expectation (suitably distant from the legacy of their more elitist and traditional roots). In particular, the anticipation of a diversified audience formed a key driver in defining what was to be displayed and how this would be done. Such reasoning manifested itself as part of an institutional awareness; that exhibition environments were now dictated more by visitor expectations of pleasure and curiosity, beyond the fact that they enabled encounters with materiality. Factors of the seemingly all-encompassing new museology – 'change, relevance, curatorial reorientation and redistribution of power' (Paddon 2014: 60, 67) – had contributed towards a loosening of curatorial control, amidst a 'shift in power towards the audience' (Lee 2007: 184).[11]

Channelling of the galleries' space

In depicting an account of a historical reality, the architectural layout of the First World War galleries was designed so as to follow a prescriptive pathway that would aid logical comprehension. The controlled (and ultimately limited) exhibition space was correspondingly moulded by the designers – allowing visitors to pause or refocus their attention, by relying upon a notion of stimulating intrigue and natural curiosity, combined with a consistency in structure. Yet this needed to be balanced with the well-founded awareness that 'the ability to pay attention is dependent on suppressing distraction' (Roppola 2012: 195). Visitor attention span continually drains as they progress through an exhibition, so a variety of delivery methods were deployed to alleviate symptoms of fatigue and boredom typically exacerbated by reading text. Principally knowledge, in the form of a textual narrative, was layered, to help facilitate conceptual understanding. This interpretive strategy was then coupled to an approach that sought to engage visitors through emotive stories, backed by material evidence. We see this 'channelling', as Roppola (2012: 213) has defined it, as forming an intricate process of organizing content in diverse and engaging ways. The function and fundamental purpose of a historical museum means that visitors require 'conceptual directedness', with the 'suitability of narrative in structuring content' (Roppola 2012: 214) widely accepted as providing a sense of overarching coherence. The narrative for these galleries' narrative was, as such, employed to impart spatially configured information. Subsidiary intentions were realised in allowing visitors to form their own views on the matters described, or to stimulate further thought. The postmodern historical museum deems discussion and debate 'vital to the development of critical engagement', and that challenging content warrants space 'for active and contemplative forms of reflection' (Macleod et al. 2015: 335). The galleries subsequently adopted a conscious mantra to actively challenge some of the dominant discourses surrounding the First World War that held currency within contemporary British society (See Wilson 2013). In so doing, they acted to convince and ultimately persuade audiences of a chosen ideological viewpoint.

The situating of text within a narrative arc was intended to both direct visitor attention and enhance understanding. Text formed one strand of an army of varying methods of interpretive media.[12] Portraying the First World War through the harmonious use of text and objects remains a constant narrative for the museum that harks back to its founding remit. In recognition of the changed backdrop of museum visitor expectations, the overarching exhibition philosophy set out to encompass a balance between delivering a clear message on the one hand and on simultaneously engaging and challenging visitors on the other. This took heed from research which had validated that many visitors 'still long for a tangible, factual and validated scholarly narrative', whilst simultaneously desiring 'more subjective information that expresses a range of differing opinions on a given topic' (Cameron 2005: 229). It was hoped, however, that visitors would not feel overwhelmed to the point of information saturation, and so the intention was to prioritise clarity and coherence over being encyclopaedic. Such a visitor-centred goal allowed for a rationed blend of clarity, mixed in with complexity, which might also leave visitors with a desire to learn more. The team of curator-historians behind these galleries consciously referred to them as being 'ground breaking' in their interpretation. This was founded on their overall approach that sought to intertwine content with asking visitors challenging philosophical questions. In their efforts to realise this, they wanted to further museological thinking about the possibilities and boundaries of design and strategies of representation. Such a culture of exhibition-making had emerged out of the more fluid version of curatorial knowledge emerging through contemporary museological practice. This, in turn, had resulted from the reworked agency of new and other stakeholders – all of whom could now influence proceedings through the multi-layered stages of the design phase.

Naturally, it is not possible to give a comprehensive account of the complexities involved in the making of these galleries within the confines of this chapter. Issues such as the process of object selection, the creation of the galleries' narrative, the integration of the Home and Fighting Fronts, as well as the philosophies and agendas inherent within these, are but a few of the subjects worthy of discussion in their own right. We have, therefore, chosen to focus specifically on one particular mantra that was applied across the whole 1100m² gallery space; that of contemporaneity. Our rationale is that this example encapsulates issues not only relating to ideas of creation and anticipated reception, but also considers the implications that this mantra had on how the exhibition space was framed.

The policy of contemporaneity

The curators of these galleries adopted a policy known within IWM as 'contemporaneity'. This set out to present historical events through the written and spoken words at the time of those directly involved and affected by the conflict, establishing it as an unfolding, undetermined drama, without any benefit of hindsight. It emerged as an approach that would prove to be hugely helpful in providing the means with which to challenge popular understanding of the War within Britain.[13] The strength of 'contemporaneity' would grant the galleries 'a directness and an unparalleled impact' – based on the premise that it was 'handing back the narration and explanation of the past to the true experts – the people who lived it' (Kavanagh 1990: 143).

'Contemporaneity' served a key role in helping the museum's curators confront the issue of chronological distance from the event that was being represented, given that the conflict had fallen from British living memory in 2009. The presentation of 'one-on-one' accounts could enhance 'the emotional involvement of visitors, through facilitating feelings of *identification*' (Roppola 2012: 237, emphasis original), as visitors can willingly project themselves into the subject depicted.[14] Museums 'are where, and other people, separated by time and space can meet. Those who work in museums make these meetings possible' (Fraser and Coulson 2012: 227). It is evident that a multitude of public history sites, at both a local and national level, have turned towards this emotionally led type of engagement; focusing on human experience as a form of empathy, but in tandem with the expected authority of an assertive organisational voice.[15] In the case of IWM, the justification of such a mantra lay partly in the fact that hindsight was a key factor in skewing visitors' existing understanding of the conflict. Embedding the authoritative yet authentic statements of historical witnesses – combined with the objects that they made, used and cared for – served as a method able to specifically confront post-event questions about the course of the conflict. Furthermore the use of the broader textual narrative retained the 'traditional role of the curator as trustworthy and 'omniscient narrator'', but in tandem with the contemporaneity policy; this allowed for the seemingly democratic inclusion of new and traditionally excluded voices.

The 'contemporaneity' policy could not, however, be applied uniformly across the entire gallery space. In terms of its creation, the concluding section known as 'War Without End' proved one of the trickiest to realise. Intent on signalling to visitors that they were now in a different world from that of the conflict, early design ideas had suggested the creation of a contemplative space reflecting a post-war chaos, suitably distinctive from preceding sections. Much debate took place regarding the date at which the exhibition would cease, with eventual settlement upon 1929.[16] Curators and designers thus had to integrate the broad consequences of the conflict, as well as examples of its impacts on a more personal level, up until this point in time. The use of intriguing objects would serve to represent a range of larger themes around the various costs of the war, displayed in a non-chronological fashion so that visitors need not move around the section in sequence. To help overcome the challenges inherent in this section, it was agreed that the contemporaneity policy could be relaxed. This was in response to visitor concentration being at its lowest ebb by this point of the galleries. The section duly featured engaging interview snippets from veterans reflecting on their experiences, an asset sourced from the museum's Sound Archive. There was additionally a tellingly reflective comment from Harry Patch, the last British Tommy to have fought in the trenches, which read 'I've tried for eighty years to forget it. But I can't'.[17] The desire to eventually permit these inclusions – and in so doing, going against a policy otherwise adhered to rigidly – proved ultimately rather revealing. Scholars have commented that an appeal to memory over history 'can imply the displacing of analysis by empathy ... memory, because of its powerful pull towards the present, and because of its affective investments, allows more readily for a certain evasion of critical distance' (Hodgkin and Radstone 2006: 8).[18] The fact that (potentially problematic) memory was incorporated in such an overt manner within the gallery space speaks more to its prominence within the field of memory studies and more broadly within the contemporary context, as 'invoked in schools, museums and mass media [to forward] political agendas which serve particular ideas about the virtues of the nation, the family or the current government' (Hodgkin and Radstone 2006: 5).

Concluding thoughts

Kjeldbaek laid down a gauntlet in asking the question, 'If a museum is primarily a collection of objects on public displays, how are they made to speak?' (2009: 363). This chapter has sought to establish a response to this, through a divulgence of agency, agendas and tacit manipulations that constitute the complex processes of exhibition-making. The museum remains a site of knowledge production, in using 'objects as elements in institutionalized stories' (Ferguson 1996: 175) that are then promoted to audiences. Furthermore the chapter has accounted for how IWM has both structured and manifested itself; our inclination is to perceive the institution's past legacy as having a fundamental influence over how it continually evolves and redefines itself in the present. As was outlined in the opening paragraphs, IWM London remains profoundly bound up in acknowledging the event which led to its founding. Distilling the complexity of the conflict in question into a spatial rendering has gone through multiple iterations during the institution's existence – meaning that it 'never stilled or settled' (Grewcock 2014: 5) as a monolithic, static site. However, as temporal distance to the conflict in question increased, the dynamic between museum makers and museum visitors shifted significantly. Curators were increasingly expected to make use of changing modes of display and interpretation to engage diversifying audiences, to allow them to consider the conflict in newly relevant and meaningful ways. As a result, when the most noteworthy commemorative milestone in its existence grew near – its centennial anniversary – IWM chose to actively re-frame its original 1917 remit against this new contextual backdrop. In so doing, it would never veer far from the rhetoric of its origins and purpose as Britain's national record to the victory and sacrifice of that conflict. Defined and nurtured by these circumstances, its inception had been 'justified in terms of remembrance and national sacrifice, of being a memorial to a hitherto unprecedented national war effort' (Brosnan 2010), and even almost a century on, any museum 'will always carry with it the legacy of its origins, for better or worse' (Barrett 2012: 112).

The overwhelming burden of education exerted on IWM equates to problematic areas of erasure and contestation, particularly the presence of imperialism within a postcolonial world, which warrant discussion beyond the scope of this chapter. It is evident, however, that the museum continues to tread a delicate line in providing a commemorative outlet that ensures future generations of Britons and international visitors are able to remember the efforts and sacrifices of the 1914–1918 wartime generation. Broad but entangled questions remain about how the politics of its displays bring it into alignment with the general mission of a war museum or memorial, and the extent to which these categories are distinguishable. Furthermore, the issue of whether a critical interpretive approach is achievable within the constraints of such a commemorative framework is worthy of future consideration.[19] What is important to acknowledge is that, as time has gone on, the criteria of honouring the extent of this national sacrifice had to be made more apparent in its relevance to increasingly distant generations. 2014 saw the institution at 'the crossroad of two approaches to reality – the museum in its creation of a new reality and the monument in its preservation of the original reality' (Lennon 1999: 74). It had to find a point of juncture between balancing an internal desire of remaining true to its origins, whilst recognising that external changes (both in museum studies and understanding of the First World War) exerted a requirement to do things differently to what had been done before. This was manifested by the institution's newfound mantra of granting its audiences an understanding of what war means and how it has shaped their lives.[20] Nevertheless, all organizations of this ilk are required to encapsulate an essentially contradictory message, that of an omnipresent factor of 'Not Forgetting' that runs deep within the institution's bloodstream, contrasted with the intentional outcome that visitors come away with an understanding of the costs of war (as bound up with the museum's original desire not to promote war in the future, whilst equally emphasizing a national sacrifice).[21] Given that visitors to IWM now bear witness to a century of conflict, visitors are seemingly inclined towards 'discursive assessments of war', that it serves purely as an inherent waste.[22] On the other hand, the metaphorical message prided by these institutions is that they enable an understanding of what war is, and such an emotional involvement can help to facilitate moral lessons. This is derived from the reasoning that insight into war experience can promote feelings of gratitude amongst visitors:

> The politics of the war museum/memorial, the manner in which it seeks to persuade, is to instil in its visitors greater understanding and respect for [those] who have served and sacrificed in war. Reciprocally, the institution serves as a site at which visitors may feel as though they *are* paying their respects, and visitors even undertake a pilgrimage to it for that express purpose.
>
> (Roppola 2012: 243, 252, 254, 259 emphasis original)

Encapsulating these roles means that it serves as something beyond just a museum; it has a stewardship role affiliated with its commemorative function (Roppola 2012: 157). IWM remains able to convey the depth of human experiences of an event that shaped Britain within the modern world. Moreover, its subject matter is bound up with the concept of identity via its engaging, emotive and raw nature. Whilst the institution continues to serve its core museological role of conserving pieces from the past, the display of these objects now functions as pieces of stimuli for commemoration. This chapter has sought to demonstrate how this was achieved within its most recent context, having accounted for influences that affected how and why particular decisions were made. IWM London now defines itself on its ability to communicate history or historical narrative to its audiences. The positive public reception garnered by these galleries adds weight to the premise that a successful formula was discovered over the course of their crafting and that the institution's curators achieved what they set out to do.

Notes

1 This chapter draws upon the doctoral research completed by Wallis between 2011 and 2014, for which Taylor acted as co-supervisor. This was an Arts and Humanities Research Council-funded Collaborative Doctoral Award that operated in partnership between IWM and the University of Exeter.
2 This institution was rebranded in 2011 as one of five 'Imperial War Museums' sites within the United Kingdom.
3 For additional commentary on the Museum's origins and early displays, see Condell (1985), Kavanagh (1994), Cundy (2015: 250–261), Charman (2008), Reynolds (2013: 209).
4 Goebel suggests that the Museum's ensuing appeal was due to the fact that 'it located family stories in bigger, more universal narratives that linked individual suffering to national and imperial survival' (2007: 167). For commentary on how the Museum's relationship between Britain and the countries of its Empire played out during this initial period, see Cooke & Jenkins (2001: 385–386).
5 Noakes has described them as 'powerful sites of cultural transmission and public education; they are an embodiment of knowledge and power' (1997: 90).
6 Museums had been 'conceived as symbols of national identity and progress, as sites of civic education for the masses' (Macdonald 1998: 9). For expansion on this notion, see Bennett (1995), Boswell & Evans (1999), Macdonald (2003).
7 For detail on this theme, consult Ferguson (1996: 178), Kaplan (1998: 37), Arnold-de Simine (2013: 8–9).

8 The point emphasized here is about the choices being made, with Kavanagh commenting that there is 'no such thing as neutral history, all acts of history making in museums are by their nature political, at the very least in what they include and exclude' (1996: xii).

9 McCall and Gray (2014: 20–21) categorise 'new museology' evolving from the perceived failings of the original museology, and the move away from museums being collections-focused and building-based.

10 As part of the social inclusion agenda under the 1997–2010 Labour Government, museums within Britain reacted to government discourses of access and equality (Sandell 2002, 2005).

11 Paddon considers this shift borne out of the commissioning of front-end and formative evaluation programmes (2014: 57). These had emerged as a result of 'pressures from social, economic and political environments', when museums were 'obliged to justify their value and worth' (2014: 67).

12 The requirement to make the galleries engaging and sensorial explains the frequent use of digital interactives and devices to deliver content in a way that sustains audience interest. It is worth highlighting the uniqueness of the historical museum in its reliance upon the interpretive tool of the written word to deliver its messages.

13 Within British public memory, the First World War equated to a 'lost generation', of poppies, poets, and heartless generals sending men needlessly to their deaths.

14 Grewcock (2014: 208) describes this type of learning as 'not restricted to an intellectual cognitive engagement but by learning as feeling, of movement and being moved'. For Fraser & Coulson, 'we are moved when we see the world through another's eyes' (2012: 231) Similarly Hooper-Greenhill observes that 'words drawn from the experience of people in real situations has a poignancy and immediacy that could not be achieved in other ways' (1994: 121). However Hanks records that 'the different small narratives of and from the people are often selected so that they add up to an uncontested account of the past' (2012: 8).

15 Andermann & Arnold-de Simine (2012: 7–8) talk of a turn towards 'personal narratives, affective engagement and imaginative investment'. Arnold-de Simine (2013: 16–17, 44–53) has analysed the deployment of empathy within international museum displays. See also Brandt (2004) for observations on how museums located on the former First World War battlefields have delivered historical interpretation.

16 This decision was based on the premise that it enabled no direct mentions of the 1929 Great Depression, and subsequent historical events. Ten years on from the Armistice was when the nation began to consider what the war had achieved, with Reynolds terming the late 1920s 'a moment ripe for reflection' (2013; 201).

17 For analysis of the problematic nature of oral records and memory, see Thomson (2013: 327–342); Hodgkin and Radstone (2006: 1–5).

18 Evans & Lunn likewise view memory as 'a process used by both individuals and more institutional forces to connect the past to the present and the future' (1997: xvii).

19 Roppola (2012: 241) defines a commemorative interpretive approach as 'distinct from a critical interpretive approach', latterly terming them 'incompatible' (2012; 259). Lennon (1999; 78) concurs, remarking that 'the [memorial] museum institution lacks the possibility of being really critical', that even choosing to commemorate 'is in itself an interpretive decision'. For Andermann and Arnold-de Simine, state-funded war museums 'perform a public role of remembrance in which they are expected to represent a broad social or at least a political consensus, producing narratives which form an integral part of national identity politics' (2012: 9). A contrast to IWM, as evidenced by its name, the Australian War Memorial (AWM) is more overt in its presence and purpose as a site of national remembrance, in acting adhesively as a site of national memory and remembrance. Worthy outlines the tensions of incorporating a public memorial role within a historical museum in New Zealand; noting the importance of

where this is located nationally determining its success of fulfilling the needs of local communities of remembrance (2004: 602, 605, 617).

20 This terminology, actively used by the reinvented Museum, achieves this by highlighting the follies and perils of warfare, whilst also demonstrating that war can be a force for positive change in some areas.

21 Essentially museums have to reconcile this inherent tension, between 'ideology, a critical view, and respect, a commemorative view. How could we help people internalise the moral lesson that we should never go through war again, while still paying tribute to those who have already served in war' (Roppola 2012: 242).

22 See Roppola's research at the AWM (2012: 242–244) for discussion on the juxtaposition between past and current conflicts, and how visitors engage in critical reflection about the nature of war in society.

Bibliography

Andermann, J. and Arnold-de Simine, S. (2012) 'Introduction: Memory, community and the new museum', *Theory, Culture & Society*, 29(1): 3–13.

Arnold-de Simine, S. (2013) *Mediating Memory in the Museum: Trauma, Empathy, Nostalgia*, Basingstoke: Palgrave Macmillan.

Ashplant, T., Dawson, G. and Roper, M. (2000) 'The politics of war memory and commemoration: Contexts, structures and dynamics', in T. Ashplant, G. Dawson and M. Roper (eds.), *The Politics of War Memory and Commemoration*, London and New York: Routledge, 3–85.

Barrett, J. (2012) *Museums and the Public Sphere*, Chichester: John Wiley & Sons.

Bennett, T. (1995) *The Birth of the Museum: History, Theory, Politics,* London: Routledge.

Black, G. (2012) *Transforming Museums in the Twenty-First Century*, Oxford and New York: Routledge.

Boswell, D. and Evans, J. (eds.) (1999) *Representing the Nation: A Reader–Histories, Heritage and Museums*, London and New York: Routledge in association with The Open University.

Brandt, S. (2004) 'The Historial de la Grande Guerre in Péronne, France: A museum at a former First World War battlefield', *Museum International*, 56(3): 46–52.

Brosnan, M. (2010) 'Putting the "M" in IWM – Incorporating museum history in the redeveloped galleries', Unpublished Internal Document, IWM Regeneration Network Drive.

Cameron, F. (2005) 'Contentiousness and Shifting Knowledge Paradigms: The Roles of History and Science Museums in Contemporary Societies', *Museum Management and Curatorship*, 20(3): 213–233.

Cameron, F. (2007) 'Moral lessons and reforming agendas: History museums, science museums, contentious topics and contemporary societies', in S. Knell, S. Macleod and S. Watson (eds.), *Museum Revolutions: How Museums Change and Are Changed*, Oxford and New York: Routledge, 330–342.

Charman, T. (2008) 'A museum of man's greatest lunatic folly: The Imperial War Museum and its commemoration of the Great War, 1917–2008', in M. Howard et al. (eds.), *A Part of History: Aspects of the British Experience of the First World War*, London: Continuum, 99–106.

Condell, D. (1985) *The Imperial War Museum, 1917–1920*, Unpublished MPhil thesis, London: University of London (held at the Imperial War Museum).

Cooke, S. and Jenkins, L. (2001) 'Discourses of regeneration in early twentieth century Britain: From Bedlam to the Imperial War Museum', *Area*, 33(4): 382–390.

Cornish, P. (2004) '"Sacred relics": Objects in the Imperial War Museum, 1917–1939', in N. Saunders (ed.), *Matters of Conflict: Material Culture, Memory and the First World War*, London: Routledge, 35–50.

Crane, S. (2000) 'Introduction: Of museums and memory', in S. Crane (ed.), *Museums and Memory*, Stanford, CA: Stanford University Press, 1–13.

Crooke, E. (2007) *Museums and Community: Ideas, Issues and Challenges*, Oxford and New York: Routledge.

Cundy, A. (2015) 'Thresholds of memory: Representing function through space and object at the Imperial War Museum, London, 1918–2014', *Museum History Journal*, 8(2): 247–268.

Evans, M. and Lunn, K. (1997) 'Preface', in M. Evans and K. Lunn (eds.), *War and Memory in the Twentieth Century*, Oxford and New York: Berg, xv–xix.

Ferguson, B. (1996) 'Exhibition rhetorics: Material speech and utter sense', in R. Greenberg, B. Ferguson and S. Nairne (eds.), *Thinking About Exhibitions*, London and New York: Routledge, 175–190.

Fraser, A. and Coulson, H. (2012) 'Incomplete stories', in S. Macleod, L. Hanks and J. Hale (eds.), *Museum Making: Narratives, Architectures, Exhibitions (Museum Meanings)*, Oxford and New York: Routledge, 223–233.

Gillen, J. (2014) 'Tourism and nation building at the War Remnants Museum in Ho Chi Minh city, Vietnam', *Annals of the Association of American Geographers*, 104(6): 1307–1321.

Goebel, S. (2007) 'Exhibitions (civic culture)', in J. Winter and J.-L. Robert (eds.), *Capital Cities at War – Paris, London, Berlin 1914–1919, Volume 2: A Cultural History*, Cambridge and New York: Cambridge University Press, 143–187.

Grewcock, D. (2014) *Doing Museology Differently*, New York and Abingdon: Routledge.

Hanks, L. (2012) 'Writing spatial stories: Textual narratives in the museum', in S. Macleod, L. Hanks and J. Hale (eds.), *Museum Making: Narratives, Architectures, Exhibitions*, Oxford and New York: Routledge, 21–33.

Hodgkin, K. and Radstone, S. (2006) 'Introduction – Contested Pasts', in K. Hodgkin and S. Radstone (eds.) *Memory, History, Nation – Contested Pasts*, New Jersey: Transactions (by arrangement with Routledge), 1–21.

Hodgkin, K. and Radstone, S. (2006) 'Patterning the National Past – Introduction' in K. Hodgkin and S. Radstone (eds.) *Memory, History, Nation – Contested Pasts*, New Jersey: Transactions (by arrangement with Routledge), 169–174.

Hooper-Greenhill, E. (1994) *Museums and Their Visitors*, London and New York: Routledge.

Hooper-Greenhill, E. (2000) *Museums and the Interpretation of Visual Culture*, London and New York: Routledge.

Jones, S. (1996) 'Making histories of war', in G. Kavanagh (ed.), *Making Histories in Museums*, London and New York: Leicester University Press, 152–162.

Kaplan, F. (1998) 'Exhibitions as communicative media', in E. Hooper-Greenhill (ed.), *Museum, Media, Message*, London and New York: Routledge, 37–58.

Kavanagh, G. (1990) *History Curatorship*, Leicester: Leicester University Press.

Kavanagh, G. (1994) *Museums and the First World War: A Social History*, London: Leicester University Press.

Kavanagh, G. (1996) 'Preface', in G. Kavanagh (ed.), *Making Histories in Museums*, London and New York: Leicester University Press, xi–xiv.

Kavanagh, G. (2000) *Dream Spaces: Memory and the Museum*, London and New York: Leicester University Press.

Kelly, L. (2007) 'Visitors and learning: Adult museum visitors' learning identities', in S. Knell, S. Macleod and S. Watson (eds.), *Museum Revolutions: How Museums Change and Are Changed*, Oxford and New York: Routledge, 276–290.

Kjeldbaek, K. (2009) 'How museums speak', in E. Kjeldbaek (ed.), *The Power of the Object: Museums and World War II*, Edinburgh: MuseumsETC, 362–392.

Lee, C. (2007) 'Reconsidering conflict in exhibition development teams', *Museum Management and Curatorship*, 22(2): 183–199.

Lennon, K. (1999) 'The memorial museum: Diluent or concentric agent of the museum institution?', *Museum Management and Curatorship*, 18(1): 73–80.

Lindauer, M. (2007) 'Critical museum pedagogy and exhibition development', in S. Knell, S. Macleod and S. Watson (eds.), *Museum Revolutions: How Museums Change and Are Changed*, Oxford and New York: Routledge, 303–314.

Lisle, D. (2006) 'Sublime lessons: Education and ambivalence in war exhibitions', *Millennium–Journal of International Studies*, 34: 841–864.

Lord, B. (2007) 'From the document to the monument: Museums and the philosophy of history', in S. Knell, S. Macleod and S. Watson (eds.), *Museum Revolutions: How Museums Change and Are Changed*, Oxford and New York: Routledge, 355–366.

Luke, T. (2002) *Museum Politics: Power Plays at the Exhibition*, Minneapolis and London: University of Minnesota Press.

Macdonald, S. (1998) 'Exhibitions of power and powers of exhibition: An introduction to the politics of display', in S. Macdonald (ed.), *The Politics of Display: Museums, Science, Culture*, London and New York: Routledge, 1–24.

Macdonald, S. (2002) *Behind the Scenes at the Science Museum*, Oxford and New York: Berg.

Macdonald, S. (2003) 'Museums, national, postnational and transcultural identities', *Museum and Society*, 1(1): 1–16.

Macleod, S. (2005) 'Introduction', in S. Macleod (ed.), *Reshaping Museum Space: Architecture, Design, Exhibitions*, Oxford and New York: Routledge, 1–5.

Macleod, S., Dodd, J. and Duncan, T. (2015) 'New museum design cultures: Harnessing the potential of design and "design thinking" in museums', *Museum Management and Curatorship*, 30(4): 314–341.

Maier-Wolthausen, C. (2009) 'The dilemma of exhibiting heroism', in E. Kjeldbaek (ed.), *The Power of the Object: Museums and World War II*, Edinburgh: MuseumsETC, 298–323.

Mason, R. (2005) 'Museums, galleries and heritage: Sites of meaning-making and communication', in G. Corsane (ed.), *Heritage, Museums and Galleries: An Introductory Reader*, Oxford and New York: Routledge, 200–214.

McCall, V. and Gray, C. (2014) 'Museums and the "new museology": Theory, practice and organisational change', *Museum Management and Curatorship*, 29(1): 19–35.

McManus, P. (2011) 'Invoking the muse: The purposes and processes of communicative action in museums', in J. Fritsch (ed.), *Museum Gallery Interpretation and Material Culture*, Abingdon: Routledge, 26–34.

Message, K. (2006) *New Museums and the Making of Culture*, Oxford and New York: Berg.

Milner, L. (2002) 'Displaying war: The changing philosophy behind the exhibitions at the Imperial War Museum in London', in H. Bronder (ed.), *Presenting the Unpresentable: Renewed Presentations in Museums of Military History*, Amsterdam: Army Museum Delft, 10–17.

Newman, A. and Mclean, F. (2006) 'The impact of museums upon identity', *International Journal of Heritage Studies*, 12(1): 49–68.

Noakes, L. (1997) 'Making histories: Experiencing the Blitz in London museums in the 1990s', in M. Evans and K. Lunn (eds.), *War and Memory in the Twentieth Century*, Oxford and New York: Berg, 89–104.

Paddon, H. (2014) *Redisplaying Museum Collections: Contemporary Display and Interpretation in British Museums*, Farnham: Ashgate.

Reynolds, D. (2013) *The Long Shadow: The Great War and the Twentieth Century*, London: Simon & Schuster Ltd.

Roppola, T. (2012) *Designing for the Museum Visitor Experience*, Abingdon and New York: Routledge.

Sandell, R. (ed.) (2002) *Museums, Society, Inequality (Museum Meanings)*, Oxford and New York: Routledge.

Sandell, R. (2005) 'Constructing and communicating equality: The social agency of museum space', in S. Macleod (ed.), *Reshaping Museum Space: Architecture, Design, Exhibitions (Museum Meanings)*, Routledge: Oxford and New York, 185–200.

Thomson, A. (2013) *Anzac Memories: Living with the Legend*, Victoria: Monash University Press.

Till, K. (2001) 'Reimagining national identity: "Chapters of life" at the German Historical Museum in Berlin', in P. Adams, S. Hoelscher and K. Till (eds.), *Textures of Place: Exploring Humanist Geographies*, Minneapolis: University of Minnesota Press, 273–299.

Trofanenko, N. (2010) 'The educational promise of public history museum exhibits', *Theory and Research in Social Education*, 38(2): 270–288.

Watson, S. (2007) 'History museums, community identities and a sense of place: Rewriting histories', in S. Knell, S. Macleod and S. Watson (eds.), *Museum Revolutions: How Museums Change and Are Changed*, Oxford: Routledge, 160–172.

Watson, S. (2010) 'Myth, memory and the senses in the Churchill Museum', in S. Dudley (ed.), *Museum Materialities: Objects, Engagements, Interpretations*, Oxford and New York: Routledge, 204–223.

Williams, P. (2007) *Memorial Museums: The Global Rush to Commemorate Atrocities*, Oxford and New York: Berg.

Wilson, R. (2013) *Cultural Heritage of the Great War in Britain*, Farnham: Ashgate.

Witcomb, A. (2003) *Re-Imagining the Museum: Beyond the Mausoleum (Museum Meanings)*, Oxford and New York: Routledge.

Worthy, S. (2004) 'Communities of Remembrance: Making Auckland's War Memorial Museum', *Journal of Contemporary History*, 39(4): 599–618.

Yeingst, W. and Bunch, L. (1997) 'Curating the recent past: The Woolworth lunch counter, Greensboro, North Carolina', in A. Henderson and A. Kaeppler (eds.), *Exhibiting Dilemmas: Issues of Representation at the Smithsonian*, Washington and London: Smithsonian Institution Press, 143–155.

8 Commemorative cartographies, citizen cartographers and WW1 community engagement

Keith D. Lilley

In these days, when fighting extends over such vast areas of country, it is necessary that not only officers, but non-commissioned officers and men as well, should receive sufficient instruction in map reading to enable them to find their way in an unknown country by a map, and grasp the general features of the country to which the map refers.

(War Office, *Notes on Map Reading* (1915), 2)

According to figures produced by the Ordnance Survey at the end of World War One (WW1), some 33 million maps were produced by the British for use both on and off the battlefield between 1914 and 1918 (Ordnance Survey 1919: 7). This extraordinary level of cartographic output is testament to the vital role that mapping played in WW1, at both strategic and tactical levels (Chasseaud 1991). A century on, the larger-scale British 'trench maps' that show in detail the landscape features of the front line, used by those fighting on the ground, are now highly sought after by collectors of wartime memorabilia, as well as reproduced by specialist publishers as facsimiles (Chasseaud 1986).

The impact of the war on British cartography was great, not only in terms of the technology of map production and distribution necessitated by the conflict, but also regarding a widening in the use of large-scale mapping by uniformed services in all theatres of the global war, as part of their everyday work behind the lines and at the front (Winterbotham 1919). The extent to which maps were used in the field by officers as well as by noncommissioned officers – for navigating the front line, for organising raids, for mining and for directing artillery – is revealed by the many annotated WW1 trench maps that survived the war, now in public and private archives and collections worldwide, which were produced through the Geographical Section General Staff (GSGS) of the War Office.[1]

Not only did the war have an impact on the production of maps but also their consumption, fostering greater familiarity with maps across a broad cross-section of British society, popularising their use more so than previously had been the case (see Heffernan 1996). Indeed, the Tommy's exposure to GSGS maps of 1:5000, 1:10,000 and 1:20,000 scales occurred through WW1, with instruction on how to read topographic maps received as part of British soldiers' basic military training and through manuals such as *Notes on Map Reading* (issued by the War Office in 1915).[2] Until 1916, maps of such large-scales would rarely have been seen, for ordinarily they were the preserve of specialists, in fields such as land surveying, estate management, civil engineering and public works. It was only after WW1 in Britain that larger-scale maps reached wider audiences and achieved greater levels of general public use, through increasing outdoor leisure activities in the 1920s and 1930s especially (Matless 1998), a trend boosted particularly by the introduction of affordable Ordnance Survey (OS) one-inch-to-one-mile 'Popular Edition' maps that featured the now well-known covers designed by Ellis Martin, begun in 1919 (Hodson 1999).

While WW1 clearly had an impact on the production and consumption of British maps and mapping, arguably giving rise to a more cartographically conversant post-war civil population of map-trained ex-servicemen, 100 years on, an interesting parallel in the mobilisation of mapping is emerging in the twenty-first century, thanks to the war's centenary commemorations. Once again, the conflict is stimulating cartographic production and consumption, though now it is through community-based centenary research projects, as well as through nation-wide WW1 commemorative projects and resources, that mapping is reaching new and wider public audiences. This connection – being forged by cartography between the war (past) and centenary (present) – can be seen in other ways too. For example, the development and application of new spatial technologies was changing maps and map-making as significantly during the war as it is today. Then it was the advance of photogrammetry and aerial photography that were having a profound technological effect on cartography, speeding up the process of revising maps and making possible rapid reconnaissance without the need for undertaking difficult and dangerous ground survey work along the front line (Winterbotham 1919; Collier 2002; Stichelbaut and Chielens 2014); now it is technologies such as remote sensing, digital spatial data and web-based mapping platforms that are profoundly reshaping not only how maps are made but also how they are engaged with and used (Laurier and Brown 2008; Crampton 2010; Kitchin et al. 2013).

Mutual innovation in cartographic methods through the development and application of new spatial technologies thus forms common ground between WW1 and its ongoing centenary commemorations. Cartographic connections between then and now, thinking through how both map production and consumption, offers a basis for understanding not only the war itself, but also provides insights on contemporary commemorative practices, and what it is to remember and memorialise the human impacts of WW1 in the twenty-first century. This chapter, first, explores these 'commemorative cartographies' that have emerged during the centenary of WW1 in the United Kingdom, using as a basis for this the proliferation of maps and mapping linked to centenary projects and programmes that are orientated towards community engagement. Secondly, it examines examples of mapping produced by 'citizen cartographers'; those individuals who are not professionally trained map-makers but who are nevertheless using cartography as a means of memorialising the war and revealing its impact on local communities.

Community engagement and the 'citizen cartographer'

One of the defining characteristics of the British commemorations of WW1 since 2014 has been the ambition by national bodies, organisations and institutions to use the centenary to engage local communities and promote interest in the war and its memorialisation. This is evidenced not least by programmes such as the Imperial War Museums' (IWM) 'Lives of the First World War' project, The National Archives' (TNA) 'Operation War Diary', Council for British Archaeology's (CBA) 'Home Front Legacy' project and the Heritage Lottery Fund's (HLF) 'First World War: Then and Now' funding programme.[3] The HLF and the Arts and Humanities Research Council (AHRC) also joined forces in 2013 to launch a joint-initiative of developing public engagement centres across the United Kingdom, to link academic and community research projects on WW1 during the centenary, giving rise to five such centres, four based at universities in England, and one in Northern Ireland.[4] Each of these nationally orientated commemorative programmes has a common purpose: of engaging local communities and a wider public in exploring different facets of WW1, particularly the war's impact on the Home Front and its effects on British population and society. The IWM and TNA projects, in particular, have used 'crowdsourcing' as a means of securing public and community inputs, drawing on a form of 'public history' that seeks to use the current and widespread popularity of genealogy and family history as a basis to help these institutions with their research and public engagement missions (see Ellis 2014). Crowdsourcing, using online and digitally based content often defines current public engagement projects led by institutions such as museums, libraries and universities, whether consciously or not, are 'creating citizen historians' as a consequence (Herbert and Estlung 2008).

The high-profile media reporting coupled to the digital resources hosted by IWM and TNA certainly captured public interest, whilst the HLF's funding programme has financed many community-based WW1 commemorative projects across the United Kingdom. There has been a certain level of scepticism by some commentators about these WW1-inspired community engagement 'public history' initiatives; suggesting that behind it all lies a under-resourced heritage sector, particularly in areas of public service in museums, libraries and archives that have seen budgets cut over the last decade (see Ridge 2014a). For others, however, such 'public engagement can involve members of the public as *citizen historians*, scientists or researchers, able to be more actively involved in academic research and to contribute to collecting and analysing data in some form' (King and Rivett 2015: 225, *my emphasis*). Forming part of the UK 'impact agenda' for universities and academics, 'citizen history' is seen potentially as a means of harnessing volunteer researcher inputs into academic projects and programmes – with collaboration and peer-to-peer training ideally – leading to mutual benefits for both academic and community researcher. This is certainly the case for the AHRC's five UK WW1 Public Engagement Centres, which were designed to connect communities of researchers, academic and non-academic alike, (and in partnership with HLF), to shape the centenary of the war through collaborative research projects that would explore its various heritage facets. While the 'citizen historian' in principle (like its equivalent 'citizen scientist') is deemed worthy by many academics (e.g. Ridge 2014b; cf. Dickinson and Bonney 2012), it has come under some criticism from others within the academy. If it is to be done competently and seriously, History (as a *discipline*, with a capital H) requires certain expertise, training and skills, for it is a specialism and a profession not simply a leisure pursuit (for discussion of these issues see Pente et al. 2015: 33–34). There remains a certain degree of ambivalence, indeed ambiguity, about the 'citizen historian', with on the one hand, a desire to see greater levels of public interest in the past, so promoting the importance of the past in the present (for both community and institutions alike), whilst a nervousness exists professionally about the kind of history being done and fears over the quality of 'the history' being produced (see Ridge 2014a). Either way, one characteristic of national commemorative projects developed with the 'citizen historian' in mind is the way that geographical information – maps and mapping, in particular – are embedded in these community engagements, and are therefore contributing to the emergence of *citizen cartographers*.

Public participation and community engagement have a place in geographical practice too, as a series of positional pieces on 'participatory historical geography' in the *Journal of Historical Geography* reveal (Bressey 2014; DeLyser 2014; Geoghegan 2014). With the growth of projects encompassing collaborative and co-production methods, linking academic geographers and the wider community, the role of 'Volunteered Geographic Information' (VGI) in geography has come under recent scrutiny (Goodchild 2007; Elwood 2008; Sui et al. 2013). Within geographical participatory research more generally, maps and mapping have important crowdsourcing roles (Dodge and Kitchin 2012). Among the methods used by geographers seeking public engagement as part of the research process are approaches based upon Geographical Information Systems (GIS), sometimes called 'Participatory GIS' (PGIS) or 'Public Participation GIS' (PPGIS) (see Dunn 2007). Here the use of mobile GPS-based web-mapping platforms for gathering and recording data about localities make use of spatial technologies in the field, as projects that rely on desktop-based online GIS systems, such as those that utilise volunteers to digitise historic maps within library collections and archives.[5] A crossover is occurring between PPGIS methodologies and 'citizen history' projects giving rise to map-minded 'citizen cartographers'. Individuals through such online initiatives are gaining cartographic skills informally, that recently were largely limited to professional cartographers and geographers.

A similar synergy occurring between geographical and historical public engagement work is evident in what Parker (2006) describes as 'community mapping', a second technique that participatory geographical research draws upon cartography. Here maps are used as a basis for encouraging public volunteers to explore their memories of places (see Lilley 2006; Milne 2016), providing a focus and a framework for *drawing* communities together. The citizen cartographer does not have to literally produce a map, therefore. The act of rereading and reimagining a place through mapping also defines the citizen cartographer. 'Remaking' the map occurs every time it is engaged with, such that the traditional perceived boundary between the 'map-maker' and 'map-user' becomes blurred. Thus, as Kitchin and Dodge (2007: 335) have argued, 'maps are constantly in a state of becoming; constantly being remade' by those who use them and engagement with them, a process of 'co-constitutive production between inscription, individual and world'. And so, rather than the map being a static object, it is fluid, and mobile. In this sense those community groups working with maps are, through their cartographic engagements, in *re*making them, becoming cartographers themselves. There are also ways of making maps literally, mapping out memories by inviting community groups to draw out their own maps of their localities, or annotate maps, to use maps to reflect on questions of place and identity – as Till (2010) does so effectively in her work with communities in post-Apartheid South Africa.

Mapping, therefore, comprises an approach that is dually creative and curatorial, since the maps themselves that arise from this process are outputs worthy of further study as cartographic artefacts. Such community mapping work offers a further context for the 'citizen cartographer' in which mapping, as a process, is iterative and didactic (Lilley 2000). To explore this in the context of WW1 community engagement, the paper examines, firstly, the development and use of web-mapping platforms linked to WW1 centenary projects in the United Kingdom and how such applications are broadening a culture of cartography within the context of community engagement projects. Secondly, it considers the way WW1 community research projects have sought to draw upon map-making as a basis for defining their projects in identifying links between localities past and present. Such aspects constitute the emergence of what may be termed the 'citizen cartographer', revealing the important roles being played by maps and mapping in shaping contemporary commemorative practices among local communities and creating legacies of the centenary of WW1.

Digital mapping and 'commemorative cartographies'

Among many WW1 centenary crowdsourcing projects aimed at cultivating and engaging the 'citizen historian' there is often a cartographic presence. At its simplest this mapping component relies on online resources, such as Google Maps and/or Google Earth – a reflection of their ubiquitous usage perhaps, of Google's domination of global mapping, and the continued and growing familiarity of its interface and functionality. Despite criticisms levelled at Google Maps, whether over its cartographic content or about ethical concerns (see Farman 2010; Brotton 2013; Garfield 2013), it is widely used and embedded into crowdsourced online projects and resources globally (Dodge and Kitchin 2012). Much of the wider critique of Google Maps, and principally its place in VGI projects, has dealt with contemporary applications for example in urban planning – rather than crowdsourcing projects that involve citizen historians. For projects dealing with the past, the use of Google Maps APIs is somewhat restrictive, as shall become clear, because they usually invite a *passive* use of digital mapping. With a little more imagination, alongside technical development and specialist input, there is scope to promote a more *active* use of web-mapping platforms for enriching and diversifying community engagement. This is achievable through more sophisticated use of available online spatial data (including historic maps as well as aerial imagery), in encouraging greater use of digital mapping technologies, particularly using Open Source (OS) geospatial data and software and publishing online content via distributed or web-served Geographical Information Systems (GIS).[6] Supporting this contention requires consideration of recent examples of web-based WW1 projects and online resources led by heritage organisations, bodies, and institutions, aimed broadly at 'citizen historians'. Reviewing what is offered cartographically as part of these centenary resources – exploring their 'commemorative cartographies' – reveals how a more analytical and interpretative use of *digitised* maps, both on- and off-line, might encourage deeper and more meaningful community engagement. In turn, this seeks to widen the role of mapping beyond simply a basis for locating centenary activity but using it for developing participatory centenary research projects.

Locating WW1 centenary activity is one of the aims of HistoryPin (www.historypin.org/en/), a web-mapping platform promoted by HLF to enable community groups with HLF-funding to 'pin' their project to a particular location, and thus share with others information about their work (as well as encourage links between projects and communities across the country).[7] As such this constitutes a successful application of web-mapping technologies in developing community partnerships, as well as serving online community centenary content, including hosting project outputs, such as photographs, and descriptions of the work itself. The HistoryPin map interface uses a Google Maps API and the functionality and familiarity that this offers. However, Google Maps content is somewhat limited and limiting, and here perhaps Open Source mapping content might have offered certain advantages, adding another level of community participation and engagement through the mapping medium beyond simply locating pins using the Google Maps API. Open Street Map (OSM) (www.openstreetmap.org) for example – apart from its perceived ethical advantages over mapping data tied to global commercial organisations such as Google or Microsoft – is 'live' spatial data, created by online communities of 'citizen cartographers'. Here lies the potential of using OSM, for example, for platforms such as HistoryPin. Rather than being a passive backdrop for locating community engagement projects, the mapping interface could itself form a part of the research process as well as an output, adding to the richness of OSM as well as HistoryPin content. With such advantages on offer, it raises interesting questions of why Google Maps are chosen in community engagement mapping interfaces such as HistoryPin, unless it is the perception that Google Maps and Google Earth 'appear more democratic and participatory than ever before' (Brotton 2013: 407). For Farman (2010: 884), however, 'the user-generated content disseminated in Google Earth by the social network [of HistoryPin] is a tool that ultimately reimagines the status of the map presented by Google and the viewer's relationship to that map', thus reinforcing the hegemony of Google Inc. itself. While this may not trouble everyone, the limited functionality of Google Maps, and its cartographic content, has been raised by professional cartographers and ultimately diminishes the potential of online mapping for community engagement and participatory research for citizen cartographers.

The scope that broadening the functionality and content of online mapping has for WW1 centenary projects is brought to the fore by a HLF-funded project led by archive, museum and archaeology services at Durham County Council, called 'Durham at War': Mapping the story of County Durham and its people in the First World War' (www.durhamatwar.org.uk/). At the heart of this online resource, aimed primarily at those living in the county in North-East England, is an interactive mapping platform. Unlike the HistoryPin platform however, Durham at War offers users ways of exploring the significance of localities in memorialising the war, and recording more about the lives of those who fought in it, as a commemorative mapping project in itself. In this sense the project uses PPGIS principles, for the mapping interface is the basis for crowdsourcing, seeking contributions from a wider public to populate the project's dataset base with information about local servicemen, in addition to helping users transcribe historical records that relate to these individuals. Rather than relying on Google Maps, 'Durham at War' provides both contemporary mapping as well as historic map content, using similar tools and functionality to those of a desktop GIS.[8] The modern and historic map layers can accordingly be navigated and overlaid, so using a transparency tool allows for juxtaposition to make the 'Durham at War' interface a research resource for users (see www.durhamatwar.org.uk/explore/).

At the same time it is a repository for locating or accessing historical information, with the added value of historic map layers that enables users to place the accounts and stories of particular individuals' lives in their geographical and historical context. The use of historic Ordnance Survey map layers are key, for the interface uses large-scale (1:2500) mapping and sheets that were revised in around 1910–11, so a date more or less contemporary with the war, as well as with one of the key sources used by community centenary research projects in the United Kingdom, the 1911 Census, now freely available online. Locating service personnel both spatially and temporally provides an interpretative context that provides an understanding of how the lives of these individuals were woven into those places they left behind during the war, though in many cases returned to. Launched a hundred years after the first Durham Light Infantry landed in France, the aim of 'Durham at War' is to gather these localised stories over the course of the centenary, as well as establish as a localised digital memorial. Key to locating lives of those who served is crowdsourcing, and here the mapping interface is used effectively as a means of navigating through the personal content uploaded by 'citizen historians'. The gathered stories about individuals can be plotted out and mapped, and from the map a pop-up with thumbnail information about these servicemen links to a fuller record with details about who they were, where they served, and with which regiment. George Patterson (1897–1916), a labourer prior to enlisting, lived at Cock & Hen cottages, Broompark in County Durham (www.durhamatwar. org.uk/story/11768/), and was killed in action on July 1 1916 at the first day of the Battle of the Somme. His is one of many lives (and deaths) recorded by Durham at War through this initiative. What is particularly compelling, however, is the scope for using the historic OS map layer to explore the local landscape that George Patterson would have known, the place he lived and worked in, and the Broompark community he was part of, which is still present. This information can orientate and link us to the particular locales of those who served, in an exercise that combines historical geography and citizen cartography.

'Durham at War' provides an interesting model of 'commemorative mapping'. However, it could be argued that the historic OS maps are used more as a back-drop to what is largely textual – archival – information gathered about uniformed service personnel, rather than using maps as a rich historical source in themselves. In the case of the entry on George Patterson, and his residence at Broompark, the 'pin' identifier is dropped onto an open field on the historic 25" (1:2500) scale OS map, rather than a particular house or a terrace. This lack of geographical preci-sion is important, in weakening the connection between the individual and the community they were part of, as well as reflecting perhaps a missed opportunity to use the map as a historical source in its own right. Perhaps mapping might be used more fully as part of the historical and geographical narrative. One way of achieving this is to make more use of the analytical opportunities maps present, either in terms of exploring spatial distribution patterns represented by the mapped data – the 'pins' on a map, for example – or in terms of interpreting the historic landscapes that maps represent – especially by combining historic and modern map layers – using web-GIS functionality; the second way to add greater depth to the narratives told about those who served is through linking maps to other external digital and online mapping resources.

Other web-mapping platforms comprising of a significant collection of WW1 topographic maps are also available for citizen researchers, including 1:10,000 GSGS3062 series of maps of the Western Front that show details of the war-torn landscapes of France and Belgium. With trench systems depicted, as well as names of trenches, redoubts, camps, and the infrastructure that was put in place by both the Allies and Germans in WW1, these online GSGS maps enable users to compare the same locality over different periods of the war. This is important for 'citizen historians' seeking to locate the places that soldiers or armies fought, with the maps revealing what the landscapes of a century ago looked like to those fighting on the ground. Moreover, locating those places where particular soldiers served in the war – globally as well as locally – is as yet an underexplored aspect of WW1 commemorative mapping projects. Making both greater and better use of available online digital mapping content would help address this, to use maps as a means of drawing connections between the Home Front and the global theatres of war that men and women saw service in 1914–18, at sea, in the air and on the ground, and reveal these spatial connections.

Community mapping and 'citizen cartographers'

The common denominator between the national centenary resources and projects led by heritage bodies and organisations, and regionally focused examples such as 'Durham at War', is their shared aim towards engaging citizen researchers, in harnessing their time as volunteers to populate databases. Primarily these crowd-sourcing projects use cartographic content as a way of helping users and contributors to navigate the content of the web-resource (beyond situating geographically the crowdsourced content). It relies on an implicit cartographic literacy on the part of the users, groups and community WW1 projects, that these resources are aimed at, an assumed familiarity with digital mapping content. The contributors are therefore, *ipso facto*, becoming cartographers through their very engagement with these mapping technologies and resources, constantly remaking maps every time they engaged with, as Kitchin and Dodge (2007) argue.

But what of the literal making of maps as part of WW1 centenary projects, of community mapping and its place in commemorative activity? Examples of works from across the UK account for community groups and projects using mapping explicitly as a medium for their studies of particular localities. This is not work being led as part of the large crowdsourcing projects mentioned earlier, but instead community-led and community generated centenary projects that are drawing together local expertise and enthusiasm for the past. This operates as a 'public history' in the sense attributed to Raphael Samuel, as a history generated from the public, about the public – from communities about communities – rather than an institutionalised version of public history, or academic 'community history', that features in some UK university History departments and undergraduate programmes (Samuel 1996; cf. Glassie 1994). Such community centenary projects have gained impetus since 2014 through funding received from the HLF, but as yet – as examples of citizen history let alone the 'citizen cartographer' – these projects have lacked systematic study as a body of work by citizen researchers. It is perhaps instructive to do so because these WW1 community mapping projects can provide a cross-section of the cartographic perceptions and practices of their researchers, at the grass-roots level, and so provide an alternative perspective to the formal and institutional commemorative mapping that forms part of the community engagement aspirations or aims of heritage bodies and organisations, and resources such as 'Durham at War'.

Two community mapping projects, each focused on WW1 and arising from centenary research, provide an interesting basis for preliminary comparison and evaluation. Both projects adopted a similar cartographic approach, yet each community group worked wholly independently of each other. Undoubtedly, many other projects have arisen through similar local community-based research, and here an examination of the content of HistoryPin might assist in undertaking a fuller exploration of WW1 community mapping. Not all centenary community WW1 mapping work belongs to projects funded by the HLF, and HistoryPin is by no means being used by community groups to record and publicise their work. Instead, it is through the ongoing activity and outreach of the "Living Legacies 1914–18" WW1 public engagement centre that a growing number of community mapping projects have been studied, and approaches of their researchers examined. The two projects examined here both relied on analogue mapping methods rather than adopting digital approaches; in itself an interesting observation bearing in mind the degree of online and digital mapping resources being provided and used by national heritage bodies for public engagement purposes. One of the longest-standing HLF-funded community research projects exploring WW1 and its effect on local communities is based in the north-east of England, and it represents an early example of a group using cartography as part of their programme. This is the Tynemouth World War One Commemoration Project (TWWOCP), which began in 2010. Its aims were 'to provide a fitting and respectful commemoration of the sacrifice of an earlier generation and a lasting record of the war and its impact on the borough', as coordinator Alan Fidler outlined in his introduction to their 2014 affiliated publication (TWWOCP 2014: 2). The project involved 70 volunteers from the locality, with the starting point for their research being a Roll of Honour for Tynemouth Borough drawn up between 1919 and 1923. This formed the basis of the creation of a database of entries relating to those named in the Roll, and accessible online for download. The database launched in June 2014 as a compilation derived from different local and national records and sources, searchable for names, birth dates, and places, and thus a biographical resource on those who fought in WW1 from Tynemouth. This digital record, and the locational data it contains on 1700 servicemen from Tynemouth, allowed the project to create a 'casualty map' displaying the distribution pattern of lives lost in WW1 in Tynemouth Borough.

Compiled by project IT leader, Steve Young, the project's map represents one of the United Kingdom's earliest examples of a community-led centenary mapping output. This is presented as a centrepiece to the project's 34-page centenary publication of 2014, reproduced in colour and labelled 'The Tynemouth Borough map showing casualties in North Shields district' (TWWOCP 2014: 16–17) (see Figure 8. 1).

As well as appearing in print, the map featured as a 'moving tribute' within some of the project's multimedia outputs, including an animated film version with narration produced by the group, and made available online.[9] Forming the basis of the Tynemouth 'casualty map' is a facsimile of a street plan for the borough that had been produced for Ward's Directory in 1915–16 by the local publishers of R. Ward and Sons. The built-up areas of North and South Shields, as well as Tynemouth, are clearly depicted in blocked shading, together with the borough administrative boundaries in red, and individual street names clearly labelled. The Ward map shows the area as it was known by those named on the borough's Roll of Honour and as a contemporary source provides an appropriate historical and geographical basis to display the disposition of casualties. There is something poignant too, in the use of a locally produced street map likely to have been seen and used by those who fought in the war, many of whom would have lived in the streets and been employed in the businesses named in Ward's Directory for which the map was created. For the project, the map narration was supplemented with information derived from the biographical database and research based on using the Roll of Honour, identifying localities of those named (northumbriaworldwarone.co.uk/database/). Thus the 'casualty map' plots out a geography of loss, street by street, neighbourhood by neighbourhood, across the borough. 'Each yellow dot', the caption explains, 'represents the location of the home of one man

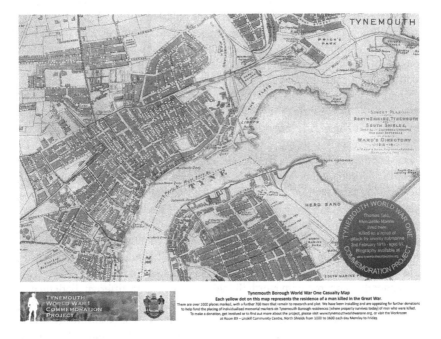

Figure 8.1 'Casualty Map' of Tynemouth Borough.

Source: Reproduced by kind permission of the Tynemouth World War One Commemoration Project.

killed during the war', a spatial distribution pattern that reveals 'the huge impact of the loss of life on the population of the borough', between 1914 and 1918 (TWWOCP 2014: 16).

The 'casualty' dots plotted out on the Ward street plan make a sobering sight. The map makes clear the streets of North Shields and Tynemouth particularly hard hit by the loss of life. Row after row of dots mirror the rows of the terraced streets shown by the map, the infilled grey of the map's street-blocks emphasising the crowded nature of the Edwardian townscape of these industrial Tyneside towns, and the cheek-by-jowl working-class existence of those who lived in them at that time. Yet there is something perhaps even more evocative and haunting conveyed by the map's serried rows of dots: their geometrical and symmetrical pattern, though clearly based upon terraces of the streets shown on the map, echo to us now the same pattern of long rows of white memorial stones, line upon line, that stand in the hundreds of British war cemeteries of the battlefields of France and Belgium where many of those same Tynemouth men fell. The 'casualty map' produced by the TWWOCP thus forms a tangible link between the lives of those who fought through linking the locality they came from with the places overseas that so many lost their lives and become memorialised in. Those volunteer researchers – or citizen cartographers – produced a map that itself is an act of remembrance, and stands as a memorial not only to those whom it seeks to commemorate but also as testimony to the dedication of the local community today. Their map of the borough is a link between past and present, an act of remembrance. With this in mind, there is an undoubted 'power' in the use of the 'casualty map', an image allowing us to orientate ourselves in time and space, in a way that perhaps other visual and creative media cannot. The map's effectiveness lies not simply in its utility and familiarity as an interpretative device, but in the potential stories that may be told through it, the possibilities for personal journeys of meditation and reflection on 'the sacrifice of an earlier generation'. In this sense, the Tynemouth casualty map is highly successful in its purpose.

Two hundred miles south of Tyneside, in the suburbs of south Oxford, a second example of a community-based centenary project has adopted mapping as a way of conveying the localised impact the war had on those who lived in the locality a century before. This project, called '66 men of Grandpont 1914–18', relates to the suburb of Oxford of the same name, an area of late Victorian and Edwardian terraced housing located around St Matthew's church in Oxford. Unlike the Tynemouth work, the Grandpont project was not HLF-funded, but relied instead on local sponsors and support, led by local resident Liz Woolley. The demographic of Grandpont compared with North Shields is clearly quite different, both then and now, yet both share, through their common approach, a desire to convey through maps the war's localised impacts, so to raise local awareness of how WW1 was brought home at the time. Like TWWOCP, the '66 Men of Grandpont' project has adopted an approach that involved different forms of commemorative activity, including a touring exhibition, a local trail around Grandpont, a film, and a book of remembrance.[10] The project map of the area produced by the project similarly occupied an important element of this work, especially the creation of the local trail that was put in place in summer 2015. According to the project information leaflet, this involved placing 'a poppy and a laminated sheet of information about each man on the gate or front wall of the house in which he lived before he went to war'.[11] The map produced by the Oxford project again likewise uses coloured dots to represent the homes of local men who saw war service, but here the approach taken is somewhat different to the Tynemouth map, for the Grandpont map has two colours, red for men who 'went to fight and died', and blue those who 'went to fight and survived' (see Figure 8.2).

The locations of their former homes were derived from detailed study of records including the 1911 Census, military service records, and the 1918 Absent Voters' list. The pattern that arises from this, on the map, reveals how many local men survived the war and returned home. Compared to the number who died in military service, the blue dots dominate the Grandpont map, reasonably uniformly distributed from street to street. This granular level of detail of the Grandpont mapping work is heightened further by the particular base map used for the exercise.

Unlike the Tynemouth 'casualty map', based on a privately produced street plan of 1915–16, the Grandpoint map uses a 1921 edition of a 1:2500 scale Ordnance Survey map. The difference in spatial scale is immediately apparent, for instead of the smaller scale blocks of shading for built-up areas of North Shields on the Ward

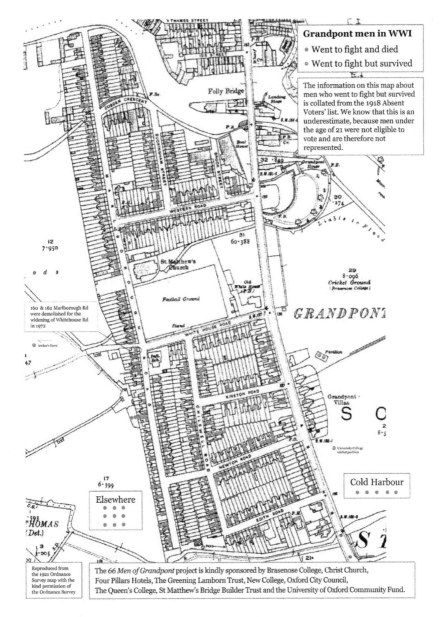

Figure 8.2 '66 Men of Grandpont' project, map of servicemen who fought from locality.
Source: Reproduced courtesy of the '66 Men of Grandpont project'.

map, the larger-scale OS map for Grandpont shows individual gardens and houses, allowing the project researchers to locate with precision the very houses where the soldiers had lived. Of course, this is helped by the Absent Voters' list, which as a source only recorded those aged over 21 and so acknowledged to under-estimate actual numbers of returnees, but included actual addresses, with which with local fieldwork could be married up with surviving buildings. This aspect of continuity in the local landscape, from 1914–18 to 2014–18, is thus used to good advantage by the project in relating the archival records to the map, and locating so precisely the homes of those who saw service. With the extra detail this affords, the Grand-pont project reveals the advantages of combining map and fieldwork, but this moreover is also put to good use through the project's aim to develop a local trail that reconnects the lives of servicemen with the buildings they left behind. This approach has a powerful local resonance, forming a link between then and now, and also between the two communities, separated by a century, that have lived in Grandpont, separated by a century. Interestingly, therefore, the map forms an important local guide to the physical and material heritage of the local townscape, and the links the project made through its research – and mapping – between the houses and the homes of soldiers. In making these connections across time and space, the mapping of Grandpont shows the importance of forging links between what is represented by the map – cartographically – and what is represented on the ground – physically – 'in the field'. Here, certainly, the combination of field, map and archival work is familiar ground to historical and cultural geographers, yet it is an exercise undertaken in south Oxford through a community-based project, linked to the local parish church, and as a very grounded and public-spirited attempt to simply record and remember in an act of local commemoration. In this sense, the Grandpont and the Tynemouth projects share much in common, likewise their approach to sharing research through a mapping medium, and placing the map centres stage as a result of this research. They both rely on groups of volunteers, enthusiasts, working not as professional historians or map-makers but as citizen historians and cartographers. What is common too is an apparent satisfaction in working with analogue mapping, not web-mapping, or using PPGIS, but paper-based maps, plotting and drawing. Neither projects are digitally averse – the data used on the maps forms part of a computerised database, and both projects have a web-presence – it is just that, as examples of community mapping and citizen cartography, the two centenary projects, situated at opposite ends of the country, have nevertheless adopted similar traditional mapping methods, and have done so to meet similar ends as exercises in local commemoration and remembrance.

The kind of localised, centenary mapping projects represented by the Tynemouth centenary project and Grandpont are not isolated example. Others can be identi-fied across the United Kingdom, similarly urban-centred. Some of these cente-nary commemorative mapping projects have arisen as a consequence of local university collaborations, such as the 'Streets of Mourning' project developed at Lancaster, and the 'Great War Dundee' project in Scotland, while others have developed through local council links, as with 'The Streets They Left Behind' project at Islington in London (see Figures 8.3).[12]

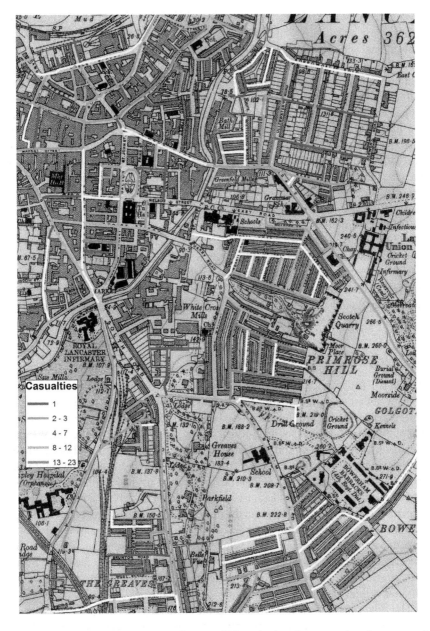

Figure 8.3 Extract from Lancaster 'Streets of Mourning' project, reproduced courtesy of Lancaster University (http://wp.lancs.ac.uk/greatwar/).

Some of these projects differ, though, in drawing on high-end technical approaches, including web-GIS and multimedia digital platforms, taking a more institutionalised approach to the mapping content. Perkins (2004: 388) similarly observed 'a great contrast between practices inside the discipline and our appreciation of mapping in the outside world'. So while in some ways the mapping outputs of the Lancaster and Islington projects are visually similar to the analogue exercises of Grandpont and Tynemouth, as examples of community mapping they are fundamentally different. Despite having some element of local involvement, these institutionally led digital mapping projects by and large represent the inputs and influences of academic researchers, rather than 'citizen cartographers'. For the latter, the mapping work itself is as much a means of building communities, of bringing local people together with a common purpose, and their map reflects this industry, literally in being a product of their labours. In this sense, these examples of WW1 community-based mapping provide a rather different, less institutionalised and academic perspective on the impacts of the war, and its commemoration. The maps resulting from these locally formed centenary projects serve to represent a spatial image of the community that creates them, reinforcing a sense of their place, geographically rooted in the local landscape, and their history, temporally linked to a 'lost' generation. The map – and the practice of mapping – forms a link between them all.

Mapping the centenary of WW1

The WW1 centenary has provided an opportunity for historical geographers to engage further with community researchers to explore mutually shared areas of cartographic interest and locality-based mapping. Resource-intense crowdsourcing approaches favoured by institutions and organisations for public engagement can be seen to have their limitations, methodologically and technically, as outlined earlier. Community-led centenary projects ought really to be able to benefit from the digital mapping content and infrastructures being put in place by heritage organisations and academic institutions up and down the land. Perhaps a way forward in this regard is to be found in the one common link between community-led centenary projects and those spearheaded by institutions: using maps and mapping as a framework for collaboration. Where this might lead remains unknown, but the map retains a power to captivate audiences, and while 'citizen cartographers' may be relatively new to the mapping scene, there is at the same time something familiar in their cartographic approach.

The dots on maps used in mapping the centenary of WW1 is common to both the projects of citizen cartographers working on grass-roots projects (in Tynemouth and Oxford, for example) as well as those projects using digital platforms and content (at Lancaster and Islington, for example). In both there are traces of an earlier paradigm where maps are seen as objective means of presenting spatial information, a form of cartography that sought to objectify the make-up of the geographical world (Crampton 2001). It is cartography that seeks stability and objectivity therefore, in a cartographic world that now speaks of fluidity and contingency (Cosgrove 1999). Such a conceptual mismatch is in itself of interest, reflecting a gap opening up between the ways maps are used and understood by many, on the one hand, a long tradition of 'scientific cartography', and on the other hand a post-processual mapping, more multi-layered, interpretative and slippery. How these two 'mapping worlds' intersect and indeed collide is an area yet to be explored by those of us interested in issues of cartographic thought and practice, yet the centenary of WW1 offers just such an opportunity for research and reflection. In so doing, it could help contribute more broadly to wider debates in the academy about public engagement practices and methodologies, as well as recent participatory geography and cartography methods.

There is something significant in how culturally embedded mapping is in contemporary commemorative practices arising through the centenary of WW1. To this end, varying ranges and levels of cartographic engagement through WW1 web-mapping infrastructure exist, with much achieved by a multitude of centenary projects to date. The spatial technologies underpinning these mirror the mapping innovations of the war itself, a symbolic link between then and now. Through new technologies of mapping, and through the commemoration of WW1, mapping literacies and audiences are evidently expanding, just as was the case through the cartographic experiences of those who fought in the war, grappling with their trench maps, aerial photographs and compasses. A hundred years on, digital mapping platforms and twenty-first-century spatial technologies offer a similar expansion of popular uses of maps and mapping for 'citizen cartographers', contributing to institutionally led and resourced WW1 heritage projects with public engagement objectives. The preliminary assessment offered here highlights how combining both approaches – the community-based methodologies of plotting on maps with the digital approaches using spatial databases and distributed GIS – has the potential to draw together community and academic researchers, using cartography as a common ground for discussion and debate. However, while crowdsourcing clearly has a role to play in these collaborative projects there is more to public engagement than developing online resources. There is a need, too, to promote their use, integrating and connecting WW1 commemorative mapping projects and resources for mutual benefit, making the role and input of the citizen cartographer key.

Notes

1 As a sample of archives and collections with WW1 British and Allied maps see: The National Archives (TNA), London: www.nationalarchives.gov.uk/help-with-your-research/research-guides/military-maps-first-world-war/ [accessed 9 November 2016]; National Library of Scotland (NLS), http://maps.nls.uk/ww1/trenches/ [accessed 9 November 2016]; 'Mapping Gallipolli', Australian War Memorial, www.awm.gov.au/exhibitions/gmaps/ [accessed 9 November 2016]; McMaster University, Canada, 'WWI Maps & Air Photos', http://library.mcmaster.ca/maps/ww1/home [accessed 9 November 2016].

2 *Notes on Map Reading for use in Army Schools*, 1915. London: HMSO. Reprinted by Naval and Military Press in association with Imperial War Museum, London.

3 Imperial War Museums' (IWM) "Lives of the First World War" project: https://livesoft-hefirstworldwar.org/ [accessed 9 November 2016]; The National Archives' (TNA) "Operation War Diary": www.operationwardiary.org/ [accessed 9 November 2016]; Council for British Archaeology's (CBA) "Home Front Legacy" project: www.home-frontlegacy.org.uk/wp/ [accessed 9 November 2016]; Heritage Lottery Fund's (HLF) "First World War: Then and Now" funding programme: www.hlf.org.uk/looking-funding/our-grant-programmes/first-world-war-then-and-now [accessed 9 November 2016].

4 See http://ww1engage.org.uk/ [accessed 9 November 2016]. This paper is based upon research undertaken as part of the AHRC-funded "Living Legacies 1914–18" WW1 public engagement centre, based at Queen's University Belfast, and the author grate-fully acknowledges here the financial support of the AHRC (Grant Ref: AH/P006671/1), as well as the opportunities for community engagement, developed through the centre's centenary activity, that have made writing this paper possible.

5 An example is the HLF-funded Cynefin project, a crowdsourcing initiative led by National Library of Wales (NLW) which invited volunteers to transcribe and geo-reference Tithe Maps within the library's collections: cynefin.archiveswales.org.uk/ [accessed 9 November 2016]

6 Such as ESRI's Story Maps platform (storymaps.arcgis.com/en/), or OS geoservers such as GeoServer (geoserver.org/) or Leaflet (leafletjs.com/).

7 The use of HistoryPin is specifically referred to by the HLF as part of its 'First World War: Then and Now' funding scheme, and a 'collection' that currently comprises (2016) a total of 2293 entries relating to HLF-funded projects is available for viewing: www.historypin.org/en/hlf/ [accessed 9 November 2016].

8 See www.durhamatwar.org.uk/ and especially www.durhamatwar.org.uk/explore/

9 The project website is www.northumbriaworldwarone.co.uk [accessed 17 February 2017]. My thanks to Alan Fidler for his comments on this part of the paper, and for allowing me to use the casualty map from the project.

10 The project website is www.southoxford.org/local-history-in-south-oxford/66-men-of-grandpont-1914-18 [accessed 23 February 2017]. My thanks to Liz Woolley for her comments on this part of the paper, and for allowing me to use the project map.

11 This passage comes from the 66 Men of Grandpont web pages: see www.southoxford. org/local-history-in-south-oxford/interesting-aspects-of-grandpont-and-south-oxford-s-history/churches/st-matthew-s-church-marlborough-road/st-matthew-s-1914-18-war-memorial [accessed 9 November 2016].

12 For 'Streets of Mourning', see www.lancaster.ac.uk/news/articles/2014/streets-of-mourning-brings-home-first-world-war-carnage/; for 'The streets they left behind', see www.mola.org.uk/projects/research-and-community/streets-they-left-behind, with web-mapping platform here: www.arcgis.com/home/webmap/viewer.html?webma p=2cb7d264996b45a6bc655c34ada3f798&extent=-0.114,51.5303,-0.0981,51.5355 [accessed 9 November 2016]; on Great War Dundee, see www.greatwardundee.com/ [accessed 9 November 2016]. Further WW1 web-mapping resources, including com-munity digital mapping projects led by the 'Living Legacies 1914–18' public engage-ment centre, can be viewed here: www.livinglegacies1914-18.ac.uk/LearningZone/ DigitalResources/ [accessed 9 November 2016].

Bibliography

Bressey, C. (2014) 'Archival interventions: Participatory research and public historical geographies', *Journal of Historical Geography*, 46: 102–104.

Brotton, J. (2013) *A History of the World in Twelve Maps*, London: Penguin.

Chasseaud, P. (1986) *Trench Maps: A Collectors Guide*, Volume 1: British Regular Series. 1:10,000 Trench Maps (GSGS 3062), Lewes: Mapbooks.

Chasseaud, P. (1991) Topography of Armageddon: A British Trench Map Atlas of the Western Front 1914–1918, Lewes: Mapbooks.

Collier, P. (2002) 'The impact on topographic mapping of developments in land and air survey: 1900–1939', *Cartography and Geographic Information Science*, 29(3): 155–174.

Cosgrove, D. (ed.) (1999) *Mappings*, London: Reaktion.

Crampton, J. (2001) 'Maps as social constructions: Power, communication and visualization', *Progress in Human Geography*, 25: 235–52.

Crampton, J. (2010) 'Cartography: Maps 2.0', *Progress in Human Geography*, 33: 91–100.

DeLyser, D. (2014) 'Towards a participatory historical geography: Archival interventions, volunteer service, and public outreach in research on early women pilots', *Journal of Historical Geography*, 46: 93–98.

Dickinson, J.L. and Bonney, R. (eds.) (2012) *Citizen Science: Participation in Environmental Research*, Ithaca, NY: Cornell University Press.

Dodge, M. and Kitchin, R. (2012) 'Crowdsourced cartography: Mapping experience and knowledge', *Environment and Planning A*, 45: 19–36.

Dunn, C.E. (2007) 'Participatory GIS: A people's GIS?', *Progress in Human Geography*, 31(5): 616–637.

Ellis, S. (2014) 'A history of collaboration, a future in crowdsourcing: Positive impacts of cooperation on British librarianship', *Libri. International Journal of Libraries and Information Studies*, 64(1): 1–10.

Elwood, S. (2008) 'Volunteered geographic information: Future research directions motivated by critical, participatory, and feminist GIS', *GeoJournal*, 72: 173–183.

Farman, J. (2010) 'Mapping the digital empire: Google Earth and the process of postmodern cartography', *New Media and Society*, 12(6): 869–888.

Garfield, S. (2013) *On the Map: A Mind-Expanding Exploration of the Way the World Looks*, London: Penguin.

Geoghegan, H. (2014) 'A new pattern for historical geography: Working with enthusiast communities and public history', *Journal of Historical Geography*, 46: 105–107.

Glassie, H. (1994) 'The practice and purpose of history', *The Journal of American History*, 81(3): 961–968.

Goodchild, M.F. (2007) 'Citizens as sensors: Web 2.0 and the volunteering of geographic information', *GeoFocus*, 7: 8–10.

Heffernan, M. (1996) 'Geography, cartography and military intelligence: The Royal Geographical Society and the First World War', *Transactions of the Institute of British Geographers*, 21: 504–533.

Herbert, J. and Estlung, K. (2008) 'Creating citizen historians', *Western Historical Quarterly*, 39: 333–341.

Hodson, Y. (1999) *Popular Maps: The Ordnance Survey Popular Edition One Inch Map of England and Wales 1919–1926*, London: Charles Close Society.

King, L. and Rivett, G. (2015) 'Engaging people in making history: Impact, public engagement and the world beyond the campus', *History Workshop Journal*, 80: 218–233.

Kitchin, R. and Dodge, M. (2007) 'Rethinking maps', *Progress in Human Geography*, 31: 1–14.

Kitchin, R., Gleeson, J. and Dodge, M. (2013) 'Unfolding mapping practices: A new epistemology for cartography', *Transactions of the Institute of British Geographers*, 38: 480–496.

Laurier, E. and Brown, B. (2008) 'Rotating maps and readers: Praxiological aspects of alignment and orientation', *Transactions of the Institute of British Geographers*, 33: 201–216.

Lilley, K.D. (2000) 'Landscape mapping and symbolic form: Drawing as a creative medium in cultural geography', in I. Cook, D. Crouch, S. Naylor and J. Ryan (eds.), *Cultural Turns/Geographical Turns*, London: Longman, 231–245.

Lilley, K.D. (2006) 'Conceptions and perceptions of urban futures in early post-war Britain: Some everyday experiences of the rebuilding of coventry, 1940–1962', in I. Boyd Whyte (ed.), *The Man-Made Future*, London: Routledge, 145–156.

Matless, D. (1998) *Landscape and Englishness*, London: Reaktion Books.

Milne, G.J. (2016) *People, Place and Power on the Nineteenth-Century Waterfront: Sailortown*, London: Palgrave Macmillan.

Ordnance Survey (1919) *A Brief Outline of the Growth of Survey Work on the Western Front*, Southampton: Ordnance Survey.

Parker, B. (2006) 'Constructing community through maps? Power and praxis in community mapping', *The Professional Geographer*, 58(4): 470–484.

Pente, E., Ward, P., Brown, M. and Sahota, H. (2015) 'The co-production of historical knowledge: Implications for the history of identities', *Identity Papers: A Journal of British and Irish Studies*, 1(1): 32–53.

Perkins, C. (2004) 'Cartography – cultures of mapping: Power in practice', *Progress in Human Geography*, 28(3): 381–391.

Ridge, M. (2014a) 'Citizen history and its discontents', IHR Digital History Seminar, 18 November 2014. Available from: http://ihrdighist.blogs.sas.ac.uk/2014/11/13/tuesday-18-november-citizen-history-and-its-discontents/ (accessed 9 November 2016).

Ridge, M. (2014b) *Crowdsourcing our Cultural Heritage*, London: Routledge.

Samuel, R. (1996) *Theatres of Memory*, Volume 1: Past and Present in Contemporary Culture, London: Verso.

Stichelbaut, B. and Chielens, P. (2014) *The Great War Seen from the Air: In Flanders Fields, 1914–1918*, New Haven, CT: Yale University Press.

Sui, D., Elwood, S. and Goodchild, M. (ed.) (2013) *Crowdsourcing Geographic Knowledge Volunteered Geographic Information (VGI) in Theory and Practice*, Dordrecht: Springer.

Till, K. (2010) *Mapping Spectral Traces* [volume and exhibitions catalogue], Blacksburg, VA: Virginia Tech College of Architecture and Urban Affairs.

TWWOCP (2014) *Tynemouth World War 1 Commemoration Project*, Tynemouth: Tynemouth World War 1 Commemoration Project.

Winterbotham, H.S.L. (1919) 'British survey on the Western Front', *The Geographical Journal*, 53(4): 253–271.

9 Affective ecologies of the post-historical present in the western front Dominion war memorials

Jeremy Foster

Introduction: contemporary tourism and the 'British' First World War memorials

Framed by a protective wall of sandbags, the centrepiece of the interpretative displays of the Vimy Ridge visitor center is a movie narrated by a young Canadian student charged with researching her country's involvement in the First World War. Depicting this history through her eyes, the movie weaves together an overview of Canadian forces' contribution to the British war effort with an account of how this military involvement in Europe socially and politically transformed society at home. This mediated narrative unfolds in the midst of the landscape where a critical battle, subsequently mythologized as a defining moment in Canadian national formation, unfolded in 1917.[1] This glimpse of the 'contemporary visitor experience' at Vimy Ridge reflects several trends unfolding across the Western Front battlefields today. Despite perceptions that we live in an era characterized by a-historical secularism and skepticism about war, interest in the First World War battlefields in Northern France continues to grow (Winter 2006; Scates 2006). From an all-time low in the early 1960s, the number of visitors to the region has grown to such an extent that the memorials constructed in the 1920s and 1930s now compete for visitors with large museums in Peronne and Ypres. Consequently, tourist-friendly media displays about the war and the site's history have become an integral part of the physical visit to the battlefields (Winter 2006).

One reason for this resurgent interest is the politics of remembrance in the home of many of these visitors. In Britain, a cult of posterity has always surrounded the war, kept alive through cultural rituals that make the war part of daily life. Britain's cities, towns, villages and schools of course have countless war memorials, but there has also been renewed interest in the media and academia about the war, and the topic has returned to school syllabi. Increased battlefield visitation has also been linked to the worldwide resurgence of family history (Wallis 2015; Winter 2012), at a time when the nuclear family is being undermined by mobility, divorce and intergenerational conflict, and cheap travel has made it easier to visit the memorials. Visits to the war graves are now regularly marketed as part of tourist itineraries, to the extent that battlefield visitation has become dependent on the infrastructure – accommodation, transportation, and information – the tourism industry provides in Northern France and Belgium. All of this has been facilitated by the growth of online media, which has dramatically increased the accessibility of information about and images of the war, and interest in visiting the battlefields. Like many 'memory places' today, the First World War memorials now come into public consciousness through web-based narratives and imagery that expands audiences, but also fragments received interpretations of the past.

While not without precedent, this intense mediated interest in the battlefields is quite recent.[2] The Imperial (later Commonwealth) War Graves Commission (hereafter WGC) had envisaged rendering the losses of a generation as meaningful through constructing two different kinds of architectural memorial on the former battlefields: dozens of small local concentration cemeteries that recalled the sacrifice of *individuals*, and a few large memorials that recalled the sacrifice of the *nation*.[3] The need for the latter only emerged a few years after Armistice, when it became evident that there would have to be sites where the ever-growing roster of 'the missing' – the names of soldiers whose bodies were never found – could be inscribed. The larger memorials also addressed the emotive and practical needs of the large groups who began visiting the battlefields from the mid-1920s, and presented an opportunity to educate future generations about the war. This decision however, created a problem: Britain's white settler colonies (or Dominions, as they had become in 1907), did not want their war dead to be subsumed into generic imperial memorials.[4] For Canada, Newfoundland, New Zealand, Australia, and South Africa, the war had been a transformative event, and political factions in each of these fledgling nations called for separate battlefield sites where the names of their own dead could be gathered.[5] This led to the construction of a handful of memorials that became de facto representational spaces for emergent colonial national rather than established British identity.[6] As a result, along with their British equivalents like Thiepval, Arras, the Menin Gate, and Tyne Cot, Dominion memorials like Vimy Ridge, Beaumont Hamel, Villers-Brettoneux and Delville Wood now trace the line of the Western Front, and have effectively become the primary destinations for battlefield tourism. This explains why, ironically, the most visited 'British' memorial today retells Canada's national formation.[7]

Articulating similarity and difference: from imperial to postcolonial memory

When completed, these British and Dominion memorials mediated a complex play of similarity and difference. All of them recorded the names of the missing, and provided a physical locale where the bereaved experienced some kind of closure. All were large edifices rather than small cemeteries, and all used the same red brick and pale limestone construction, together with a restrained Beaux Arts classicism that seamlessly accommodated the standardized a-religious sculptural iconography that was deemed appropriate, and the stone panels recording the names of the missing.[8] All these memorials included spaces that can accommodate a gathering of thousands, often aligned with a vertical architectural element visible from afar that, while not overtly religious, suggested imaginary connections between earth and heaven. In effect, a transportable set of architectural principles encoding the cultural-symbolic values of the British Empire – for example, a large axial configuration focused on an eye-catching architectural feature – was adapted to local site circumstances.

While the construction of these larger memorials undermined the WGC's original intentions of confining commemoration to modest cemeteries, once completed, it elaborated the topo-centric remembrance that was so potent in the smaller sites.[9] Because of the nature of the hostilities in this theater of the war, all the Western Front memorials are located on landscape eminences, which are rare in Northern France (Johnson 1919: 517). By simultaneously mapping colonial troop's 'actions under circumstance' in relation to the geography of the Front and the actions of their European counterparts, the Dominion memorials reinforced the WGC's topographical commemoration, but gave it a subtly national symbolic value. There were, however, differences between the British and Dominion memorials as well as the Dominion memorials themselves, in terms of the size, configuration and iconography.[10] These differences reflected the nature of each nation's participation in the war, the narratives, attributes and iconography each used to record that participation, and the cultural and political debates generated at home by the memorial's construction.[11]

Today, most visitors initially see little difference between the British and Dominion memorials. They all commemorate the same war, and they all use a version of the same architectural vocabulary found in many other large civic buildings of the period. This Classical 'high diction' was combined with high quality craftsmanship and materials, and in some instances – especially sites designed by Edwin Lutyens – became abstracted to a degree that it took on an elemental, 'timeless' quality that was subsequently seen as a signature strand in 20th C British semiotics (Geurst 2010: 183; Worpole 2003: 164). Furthermore, as in the cemeteries, the orchestration of nature was crucial. Despite original calls to keep the devastated landscape as it was,[12] the same landscape design vocabulary prevailed – a combination of manicured green lawns, long-lived shade trees, evergreen shrubbery, masonry walls and carefully tended flowering perennial beds, with its connotations of the peaceful churchyard or 'improved' estate of rural England (Morris 1997; Keegan 1997; Miller 2012).

This does not mean, however, that over the course of their lifespan, the function of the large Western Front memorials as *lieux de mémoire* has not reflected the shifting social and political interpretations of the war in each of the nations concerned.[13] These shifts have been most significant in the ex-Dominion memorials, as Canada, Australia and South Africa drifted away from Britain, and developed their own multicultural, multiracial and postcolonial identity.[14] This diverging national-representational use of these memorials originated in the debates surrounding each memorial's funding and construction, and continued with how each Dominion sought to control their memorial's management. With the exception of Canada, these nations have mostly left daily oversight and maintenance to the WGC. Modifications to the original architecture have been extremely minimal, focused on maintenance issues (the youngest of these memorials is now nearly

80 years old)[15] and accommodating changing patterns, practices and volumes of visitation, which are placing new strains on some Western Front sites.[16] Changes to these memorials' role as spaces of representation have largely been limited to amending their interpretive displays.[17]

Recently, however, as each former Dominion prepares for its centenary, this policy of non-intervention has been replaced by a more engaged attitude towards their memorials.[18] Each of these countries is using this anniversary to mobilize citizenry and funding at home for projects that expand the interpretation of their participation in the war.[19] In the summer of 2017, a new permanent education center, funded by the Canadian government and the Vimy Foundation, will open at Vimy Ridge. Also slated to open in 2017 is a new visitor center at Villers-Bretonneux, named after General John Monash – a formalization of the dawn services held at the site since 2000, after decades when the largest overseas Anzac Day services occurred at the Dardanelles landing sites. Meanwhile, at Delville Wood, 20 years after the end of apartheid, at an annual memorial service during which African soldier was reinterred, South Africa's deputy president promised to modify its museum to reflect the contributions of black South Africans to the war. In 2016, a new panel listing those drowned in the 1916 *SS Fendi* sinking was scheduled to be unveiled at the Delville Wood.[20]

This recuperation of the national-representational function of the large Western Front memorials is probably inevitable, given the onset of the centenary, and the need to compete with more contemporary battlefield destinations. But it is also about the politics of the present – the rewriting the nation's history in order to appeal to the current generation's sensibilities – and therefore needs to strike a balance between asserting an abstract idea of national belonging and avoiding dated or crude appeals to identity (Mycock et al. 2014). As postcolonial theory's questioning of cultural essences or meanings recognizes, identities are continually reconstructed through social practice and political discourse. Just how far notions of 'national identity' have evolved after decades of cultural hybridity is demonstrated by the fact that the young Canadian doing the remembering and identifying in the Vimy Ridge movie is ethnically Asian and female.

Battlefield visitation: motivations, practices, understandings

This reminds us that even while the memorials themselves are physically little changed, the *nature of remembering* occurring at them has changed, reflecting visitor demographics and expectations and the practices through which they encounter the memorial. There are now multiple motivations for visiting, just as there are different audiences and reasons for doing so.[21] Visitors are now three or four generations removed from anyone who participated in the war, and live in a 'post-national' era when many are thought to be embarrassed by overtly formal expressions of nationalism. Although about half are British or French, some are German; the balance come from an array of other countries, including those not originally involved in the conflict. A few are there to participate in official ceremonies, others to trace threads of family history; most, however, are simply tourists, some but not all of whom are visiting the memorials as an educational or self-improving activity (see Figure 9.1). Other, overlooked visitors are local dog walkers, runners, farmers, and environmentalists who also use these sites regularly. Not only composition and motivations of people found at these sites, but also the ebb and flow of visitation, shifts from day to day, and season to season.

Figure 9.1 'Learning about who we are'; Vimy Ridge, 2014.

In one respect, contemporary visitation does perpetuate the original function of these memorials: a significant number of visitors still arrive in organized groups.[22] In the past, these groups were participating in pilgrimages', today they are likely to be part of commercial tours, school groups and other 'elective communities' that arise in contemporary life, for whom 'team building' is an important motivation. Because these groups are looking for an uplifting, collective experience, they naturally gravitate towards the large memorials. Historically, pilgrimage has been seen as distinct way of construing a 'memory place'. Unlike tourism, it implies arduous journeys and is animated by the idea that personally visiting a place transcends other ways of incorporating spiritual or cultural beliefs into one's own life. A pilgrimage is made collectively, by individuals whose differences are subsumed by the desire to visit the site, and the bonds that develop during it are further heightened by the often cathartic encounter with the site (see Figure 9.2). Finally, those who make the pilgrimage acquire heightened status in the society that deems the site significant. Although few contemporary groups have overtly nationalistic agendas, or undertake arduous journeys, they recuperate some of these large memorials original function as imaginary poles of existence, or 'centers out there', simultaneously removed from yet making sense of everyday life (Turner 1974).

Figure 9.2 Team building and catharsis; a group of Canadians cycles past Vimy Ridge memorial.

Source: Adam Day, Legion Magazine www.legionmagazine.com, 2012.

Although visitation to these large memorials now has touristic overtones, whether it occurs individually or as part of a group, remembrance still lies at its core (Scates 2006). However, this reveals relatively little about *what* is remembered, or *how* or *why* this is remembered during these visits. Monuments and memorials are situated within continually shifting fields of signification. As the original event memorialized becomes less the property of those who were there, it is either transformed into a symbol of larger ideals (such as 'national identity', 'universal brotherhood', the 'waste of war') or simply forgotten. Alternatively, it may be remembered differently, as succeeding generations bring other life experiences from those who created the memorial, let alone those remembered in them. This increasing destabilization of the relationship between artefacts and remembrance by the passing of time was originally recognized by Alois Riegl in his distinction between *intentional* and *unintentional* monuments (Riegl 2004: 57–58). Both kinds of monuments are characterized by commemorative value; however, in the former (which Riegl associated with 'historical value'), this value is determined by the *creators* of the monument, while in the latter (which he associated with 'age value'), value is determined by *contemporary users*.

This raises questions as to what kinds of meanings the large Western Front memorials evoke in visitors today. How accessible is the hybrid identity they originally strove to project – simultaneously national and yet part of a supranational,

Anglophone community? Alternatively, what kinds of remembrance might they evoke today, and for whom? And, relatedly, why are some of them more popular than others? Surveys show that beyond teaching them about the war, and allowing them to (in the words of one visitor) see "an important part of who we are", most contemporary visitors find these large memorials extraordinarily moving, even life changing. This effect seems to have less do with the iconography embedded in the memorials than with the feelings provoked by 'being there'. Indeed, its seems that in these memorials, the selective organization of 'the memorable' through architectural design is continually upstaged by the lived workings of memory. Here, it is helpful to recall Stanford Anderson's distinction between 'memory *through* architecture' – the intentional use of designed environments to trigger collective memory – versus 'memory *in* architecture' – the unintentional material and temporal relations mediated by these environment (Anderson 1995).[23] Tellingly, the First World War memorials have been identified as one example in architectural history when "the former makes unusual demands on the latter" (Lipstadt 1999: 66).

But what then, might the 'lived workings of memory' be in these memorials? And how do they differ from those awakened by seeing images of these designed places in a book, museum or online, say, or by watching an interpretive movie about the war? One answer to this might be found in frequent visitors' comments about "the peaceful atmosphere of the surroundings" and the memorials relationship to "one of the most subtle landscapes I've ever visited".[24] Crucially, all the large battlefield memorials are as much *landscape*s as they are *architectural edifices*. This goes beyond the fact that they encompass extensive reconstructed ground, walks, lawns, and woodlands, and are situated far from built-up areas. When built, these memorials also embodied a yearning for a harmony between humans and natural world, and a coherent, meaningful place that facilitated a process of identification. This process was, and continues to be, facilitated by the uniquely multiplicitous stories landscapes can tell: not just the story of their making and the story they set out to tell, but the story (or stories) they invite us to attach to them.

The narrativity, latent in all human artefacts, is particularly strong in cultural landscapes, whose physical vulnerability in the face of passing time lends them a fundamentally elegiac tonality. In the Anglo-European world at least, this narrativity is, paradoxically, augmented by the association of nature with the 'consolations of persistence', and the way the pastoral articulation of nature encodes a rich spectrum of myths, memories and obsessions.[25] Cultural landscapes can, depending on how they are constructed and managed, mediate decay or renewal. They can suggest either poignant analogies between culture and cultivation, human life and fragile nature, or, alternatively, the consolations of cyclical regeneration afforded by the non-purposive agency of natural processes.[26] This inherent narrativity in the pastoral imagination is already implicit in Riegl's categorization of monuments. According to Riegl, the *intentional* monument manifested a desire to preserve the artefact's appearance unchanged and avoid natural processes of decay, the *unintentional* – the idea that natural traces left by the passage of time are in some way generative nd producing aesthetic effects.

This suggests that the appeal of the large – that is *landscape* – memorials rests as much on how they engage the passage of natural time as it does on embedded historical narratives; in other words, in their transformation into unintentional monuments. Arguably, due to the vagaries of history, this transformation has been particularly potent in the Dominion memorials. Many who now visit them come from different parts of the world than those memorialized in them, or who created them; although most visitors recognize the distance separating the memorial from that home, they know little about the imagined communities they were designed to represent.

Landscape as memory-theater: immersion, affect, temporality

Thinking with (rather than *about*) landscape in this way highlights the medium's potential as a "theater capable of [. . .] bringing the past into the present". Landscape-as-theater is landscape "encountered, sensed, and *performed*"; it is ultimately through the two-way dialogue between landscape and subject that "memory is activated and the past becomes a substantive part of everyday life" (Della Dora 2008: 229). This mnemonic activation relies on how the experiencing subject – not just present but *moving* through the landscape, open to corporeal sensation as well as material relations mediated by it – simultaneously draws on the history of ideas, references, and worlds implicated in that landscape's creation. Thus, 'landscape stories' are precisely ones in which lived understandings and other-than-lived associations are syncretically entangled.[27]

Our conceptual understanding of this immersive articulation of landscape subjects and objects has been expanded by the recent 'affective turn'. Understood as forms of pre-reflective emotional intensity and the rising and falling of the 'lived power or potential to act', affect has also been described as the syncretic constitution of our sense of ourselves in relation to context. It is a matter of agential relationality constantly evolving in lived/living space and time through the intersection of the potentialities of objects and bodies; it *emerges through seemingly involuntary embodiments and interactions between and with present objects and environments* (Shields 2011: 120, emphasis added). These rising and falling potentialities play out through "rhythms and refrains" that, reaching a point of expressivity, constitute a particular ambience or atmosphere (Hunter 2015: 189). Because these potentialities are also *modifications of states over time*, affect is therefore associated with liminal transitions, and emotional dispositions become synonymous with quasi-temporal orientations (or *displacements*) towards 'other (possible) worlds'. The three most common of these are '*nostalgia*' (oriented towards the past), *desire* (oriented to the present) and '*hope*' (oriented towards the future). Different places engender, or are 'haunted by', different configurations, or *ecologies*, of these and other affective modalities, in which tensions appear without necessarily 'making sense' (Park et al. 2011: 8). Geographically, the affective quality of a place or setting is linked to the emotional color of potential embodiments and relationalities it awakens.

Understanding the affective potentialities of the large battlefield memorials thus requires a triple focus: on the emotional states and bodily dispositions of visitors; on the specific embodied interactions through which these are modified; and on the performativity of these embodiments – their spacing, duration, and cadence – that help move visitors both literally and *empathetically*, into the site. The '(memory)-place' becomes as much an ecology of relational animations as it is the stage on which these to play out; the dominant atmosphere is a matter of the performative interactions – between bodies, between bodies and environment, and in the environment itself – that blur subject-object distinctions in the lived present. As Henri Lefebvre's concept of rhythmanalysis emphasized, not only the ongoing embodiments in any locale but also the *temporalities* are multiple – simultaneously steady, intermittent, volatile *and* surging. Thus, locales are always both in the process of becoming, and stabilized (Lefebvre 2004). Temporal ebbs and flows are also implicit in spectrality: the sense of the co-existence of the already-passed and the not yet, disrupting relationships assumed to occur according to linear temporality and meta-narratives of history, but also opening our eyes to the agency of presences that remain (Edensor 2005: 189). These kinds of tensions are especially vivid in memorials which, as they lose intentionality, allow earlier inhabitations, and ways of thinking and acting to be seen in new light, inflecting social experiences and material understandings in unforeseen ways.

This hermeneutics of temporal embodiment resonates with what Dipesh Chakrabarty (2007) calls *affective histories* that "find thought intimately tied to particular places and forms of life". He attributes this quasi-Heideggerian re-attunement to the diversity of human lifeworlds to the crisis of temporal thinking brought on by Europe's displacement from the center of the master narrative of world history (Chakrabarty 2007: 18). This can be linked to Fredric Jameson's argument that late capitalism's political disorientation is linked to the waning of the sense of *duree* that previously supported master narratives of history (Jameson 1991), something that has only increased as neo-liberal capitalism's 'perennial gale' of creative destruction (manifested in other such nation-undermining effects as religious fundamentalism, mass immigration, and climate change) spreads globally. This has led some cultural theorists to talk of the 'end of history', and a disruption of settled relations between the past, present and future (Harootunian 2007). The wager here is that in an era of not only postcolonial but also post-national and post-historical time-consciousness, remembrance becomes increasingly local and phenomenal, introducing new potential articulations between embodiment and history.

From this perspective, compelling arguments could be made that all the large Western Front memorials, once harbingers of reconstruction, now act as windows, not so much into the past, but into a disappearing time-consciousness. However, if this is so, it is precisely the *Dominion* memorials – originally highly idealistic expressions of the meta-narrative of Eurocentric history – that become the most potent 'echo-chambers', landscape theaters in which worldly, embodied understandings awakened by the passing of time are most affect-ive.

The affective work(ings) of landscape: spatio-temporal enactments and material latencies

Reconceptualizing these landscape memorials as "unfolding . . . fields of relations that [cut] across the emergent interface between organism and environment" (Ingold 1993: 156) underlines that their affects are implicitly kinaesthetic, linked to how bodies disposed for particular modes of action experience modifications *arising from relations between ideas* (Whatmore and Hinchliffe 2010: 446). Connections between situated bodily movement, understanding and emotion has long been recognized in landscape design (Conan 2003); it also lies at the heart of readings of the war memorials as dramaturgical places, in which individual feelings are produced by the collective, usually socially constructed, space ballets (Gough 1998). Both relationalities are captured in the argument that at these large memorials "movement is not . . . encouraged for its own sake, but to draw [the visitor . . .] from afar, and to encourage them to penetrate its immense depths"— see Figure 9.3 (Lipstadt 1999: 68). As we have seen, some version of this obligatory *promenade architecturale* is an integral part of all the large memorials. But again, this feature has unique resonance in the Dominion memorials, which were originally created to mark (and frame) a single, heightened spatio-temporal event. Their meaning as memory places was related to their function as the terminus of a ritualistic once-in-a-lifetime journey halfway round the world, rather than the rhythmic continuities of everyday life (Foster 2004).

Consideration of 'embodied practices' shifts attention to the temporal rhythms (linear, sequential, cyclical) that structure the kinetic event of visiting the memorial. Most obvious of these are the temporal rhythms of visitation:

Figure 9.3 'Movement not for its own sake, but to draw visitors from afar'; Villers-Brettoneux, 2014.

the irregular rhythms of individuals visitors and tour groups, tied to 'everyday' practicalities like school holidays, and the more scripted, pilgrimage-like rhythms determined by key 'historical' anniversaries, with their political speeches, rhetorical benedictions, and fleets of dignitaries. Set contrapuntally against these rhythms (brief intensity interspersed by long intervals of emptiness) are the recurring rhythms of local uses, and the almost imperceptible 'natural' rhythms of the seasons, weathering and aging that require cycles of remaking of the memorial that themselves become part of the site's life. In fact, practices involved in visiting these memorials have always been curiously entangled with those involved in sustaining their material existence. Precisely because the original event through which these places acquired meaning (i.e. the battle) was so fleeting, they had to remain constant, unchanging and physical manifestations of the names they acquired during the war and which became inscribed in colonial national cultures through popular discourse. This elision of narrative and emergence marked these memorials as places that existed in *different time-space* from their surroundings.[28]

This entanglement of the sites' physical architecture and social choreographing of corporeal practices still resonates today. When confronted by the carved stone panels, one becomes aware of the sheer number and individuality of those killed, and the poignant distance between their places of birth and death, rather than the heroic narratives of nationhood. But, as in the past, this corporeal encounter is often also collective; enacted as a part of a larger kinetic 'event' whose brevity and intensity contributes to its emotive resonance. Visitors are 'placed' in some kind of empathetic solidarity with both the original pilgrim (for whom the site was designed) and the original soldiers (who fought over this particular piece of topography).[29] A collective, quasi-reverent inhabitation of the sites, along with their Beaux Arts axiality, induces an involuntary rituality in how visitors approach and navigate the site, and comport themselves in relation to its spaces and structures (see Figure 9.4). Cutting across these empathetic relations, however, is the fact that, compared to the original pilgrimage participants, today's visitors have very different life experiences, and less personal relationship to the memorial. This encourages an oscillation between a reflective inward-looking personal subjectivity,[30] and a more intersubjective, outward social awareness of how individuals become entangled in the grand narratives that shape world history.[31] The fleeting but embodied visit prompts involuntary solidarity with other, present and past, visitors, promoting cross-generational insights and transnational affiliations that cut across historical time (see Figure 9.5).[32] This is enhanced by the way private, subjective temporality of the visitor experience and the national, historical temporality of the ceremonial remembrance each leave traces in the landscape for the other to find.

Equally important to the large memorials' affects is their materiality. By this I mean not their architectural vocabulary (whose original retrograde connotations are now forgotten) but rather the "the displacement of elements in their contexts" through the "procedures of a technical nature" involved in their making and maintenance (Berrizbeitia and Pollak 1999: 48). Materiality also

Figure 9.4 Collective quasi-reverential enactment of the memorial; Vimy Ridge, 2014.

encompasses Anderson's "unintentional material and temporal relations medi-ated by the environment", and how "appearance and apportionment" becomes "conjoined with utility and strategy" (Corner 1999: 49). Although these memo-rials started out as assemblages of carefully crafted material elements and symbolic iconographies from other places, over time they have developed their own native material semiotics and rhythms—see Figure 9.6. Thus, while all memorials' affects are predicated on the refusal of change, this is especially true in those on the Western Front. A memorial like Vimy Ridge may be built out of Croatian limestone chosen for its extreme durability, but the way its mass adapts itself to the earth also seems to glorify the dead in a more durable way than a mere stone edifice (O'Shea 1997: 92). These material semiotics are also a matter of management practices; maintaining these sites involves the regular upkeep and occasional replacement of their architectural as well as their natural elements. While some of the original elements have been lost, enough survive to mediate the sense of a world that has not only been made but also *repeatedly remade in exactly the same way*, defying not only entropy but also transforma-tion.[33] Paradoxically, this ongoing renewal of the landscape, quite literally, *keeps remembrance alive*.

Additionally, because they were built in open countryside, the 'third nature' (i.e. nature worked for symbolic purposes) of these large memorials has always been in an active dialogue with the 'second nature' (i.e. nature turned to productive ends) of their surroundings.[34] As countless historic images show, these islands of civilization were initially conjured out of a churned up wasteland of mud, debris and corpses, a destroyed condition now sometimes called 'fourth nature'.

Figure 9.5 Disrupting relationships that occur according to linear temporality; Delville Wood, 2014.

Today, traces of that human devastation can still be found, though most of it has been obscured by restoration; the former battlefields are now as intensively cultivated as before the war. While this has helped the surrounding fields, woods and villages regain something of their original character (Heyde 2015: 194),[35] the Western Front remains a region where nothing is more than 80 years old, and most things have been reconstructed quickly and cheaply (O'Shea 1997: 37). The experiential juxtaposition of the timeless landscape memorial and the everyday surroundings reverses the memorials' ability to draw visitors towards themselves, and provokes an expansive, yet temporally unsettling reverie.[36] The memorial's unchanging, 'artificial' third nature becomes paradoxically spectral compared to the mutating 'second nature' of the surroundings, in which the passage of time has merged with topography. The more these surroundings evolve, the more the designed memorial feels like it belongs to the self-regenerating, pastoral rhythms of the natural world rather than the disorderly power lines, motorways, and wind farms that have begun to encroach. This otherworldy 'naturalization' of the memorial is further heightened by the seemingly anachronistic use of sheep to maintain the grass at some of these sites.[37]

Of course, none of these material affects can be isolated from other symbolic meanings and cultural representations that shape structures of feeling towards these memorials. Today, as we have seen, images of past people and places heighten rather than detract from a visitor's encounter with physical monuments; the interaction between these two registers of place signification work together to create a 'world' that seems to have always been in circulation. This is now being further augmented by various digital technologies that change the experience of the memorial, because they place visitors in the midst of a multiplicity of narratives, both near and far, in time and space.[38] These spatio-temporal displacements (or, *emotional re-orientations*) by other-than-lived associations can be pungently antinomial. As one author who was brought up to disregard the past tellingly recounts,

> When visiting the battlefields (I realized) the War took place in color, *not in the black and white photos in history books*; that this wet sponge was their sky, this sweet bird song was the same as the one [soldiers] heard in their muddy ditches.

> (O'Shea 1997: 37, emphasis added)

This promiscuous crowding in of material insights, however, could be prompted by all kinds of after-images; for instance, the letters, poems and memoirs in which soldiers emphasized the beauties of nature that somehow survived in the landscape, and which compensated for the appalling destruction all around. These literary records show, alternatively, that this beauty could intensify sorrow, because it triggered the realization that this pastoral landscape was being destroyed. These kinds of understanding become especially poignant for those hailing from the original Dominions, whose soldiers came from less bucolic environments, and often saw the European landscape as a pastoral home.[39]

Figure 9.6 The facts of history haunted by non-purposeful effects of natural time; Thiepval, 2014.

Conclusions: the passage of time and post-historical affects

As in other projects that seek to bury the past, the commemorative strategies used in the Western Front memorials sought to redeem a haunted place. For most visitors today, however, these memorials do this not by resurrecting historical facts, but through presences and understandings that simultaneously narrate and disrupt the telling of those facts. They have, in effect, become monuments to themselves, even more than monuments to the events or individuals they nominally commemorate. They are now 'memory places' not only for the catastrophic events of 100 years ago, but also the belief in the importance of spaces that escape the utilitarian logics of the production, consumption, or spectacle. As such, they offer a sense of orientation in the face of historical decline and a loss of moral certainty, and a redemptive sense of existential belonging that has itself begun to seem anachronistic today. And the longer these memorials resist the passage of time, the more they heighten awareness of it, helping to reproduce the frisson in each successive generation of visitors that theirs is the last for whom the war has any significance (O'Shea 1997: 80).

Crucially, this otherworldly quality stems from spectrum of relational understandings awakened by physically visiting these large memorials. These include their now-modest scale; the general air of sanctity imparted by their classical iconography and limestone luminosity, enlivened by the shifting cloudscapes of the region's maritime climate; the contingent, quasi-ritualistic and collective enactment of their spaces; the cross-generational sympathy visitors feel with the bereaved; the resolute control of the lives of their natural and man-made elements across decades. These relationalities are only indirectly related to the commemorative agendas of the supranational Anglophone community, or the meta-narratives of nationhood that brought many of the soldiers to this place. *In situ*, however, they combine to create a pervasive nostalgic mood that supplements the spectral presence of those who died here, thousands of miles from home. The cognitive character of the nostalgia the memorial induces will vary; depending on the visitor's origins; they could invite recall of a once powerful imperial nation's benign relationship with its colonies, or a longing for a 'simpler', less 'multi-cultural' state of nationhood. However, such nostalgic orientations will always depend on how the *ontologically elegiac* materiality of the landscape invokes the *absent* imagined community and 'world' of spectralized European history.

This transformation of these memorials into repositories of past dreams and hoped-for futures is heightened by other museological representations of the war that play down national difference in order to promote European integration (Cadot 2014), as well as the growing post-historical sense that Europe is no longer the author or measure of global time-consciousness. It is above all in the Dominion memorials that the complex of reifications through which the meta-fictions of European history become fixed in space are most poignantly exposed. Through a flickering kaleidoscope of relationalities in the lived present, the ghosts of those who died in these locales have given rise to a plenum of 'almost-there-ness' that overwhelms these landscapes with affective intensity. These memorials are not only sites deeded in perpetuity by one European nation to another in order to memorialize the actions of a third non-European nation, they are also non-productive landscapes where, paradoxically, no one has ever belonged, yet are actively maintained to represent a highly cultural yearning for the pastoral. The contemporary entanglement of everyday, historical and natural temporalities in these memorials extends this original spatial and temporal distantiation from the surroundings, contributing to their affective ecology. Thanks to a shift from *historical* value to *age* value, the non-purposive effects of natural time and the received facts of history now haunt each the other in these Dominion sites. Instead of national identity, these memorials now mediate an *unhomely* sense that cultures are continually subject to impersonal forces, and exist in multiple time frames.

Arguably, then, the Dominion First World War memorials would seem to epitomize arguments that perhaps "place takes place through a spectral event of displacing" (Wylie 2007: 180–181). The quasi-temporal displacements at work here are multiple, multi-scalar and ultimately, incommensurate. This creates a powerful affective ecology that, disrupting conventional articulations of past and present, landscape and geography, embodiment and history, uncover the ontologies whereby the identities of people and place are co-constructed. By combining retentions from the past and protensions for the future that are both resonant *and* dissonant, the Dominion memorials are now not just unintentional monuments. They are highly emotive places of becoming, where historical and post-historical time-consciousness brush against each other, drawing, awakening and orienting ever more layers of nostalgia, desire and hope.

Notes

1 Vimy Ridge was the first time all four of Canada's divisions fought together under their own general; the taking of the Ridge, which overlooks the strategic Douai plain, was militarily critical.
2 On the early wealth of visual imagery about the battlefields, see for instance Heyde (2015: 184–187).
3 On the history of the WGC, founded by Royal Charter in 1917, see Gibson and Kingsley Ward (1989); also www.cwgc.org.
4 At the 1907 Colonial Conference, Australia, Canada, New Zealand and South Africa were accorded 'self-governing Dominion' status. Their participation in the War made them 'minor belligerent powers' at Versailles.
5 Indeed, it was the experience of the War that made some in the Dominions want to distance themselves from Britain. See O'Shea (1997).
6 Eventually, a total of 12 large 'British' memorials were constructed in France and Belgium.
7 Since 2011, Vimy Ridge has received about 700,000 visitors every year, Thiepval around 150,000. (Sources: Vimy Ridge: visitor records provided by Maj. Guy Turpin, CD, Commemorative Sites European Operations, Veteran Affairs Canada, July 2014; Thiepval: Winter (2012).
8 Exceptionally, South Africa's names were incorporated at Thiepval and the Menin Gate rather than at Delville Wood.
9 This avoidance of overt nationalistic expression had several causes. The WGC wanted to play down differences between British and Empire soldiers, avoid reminding visitors of the bellicose jingoism that had shaped the War, and respect France's wishes that none of these 'foreign monuments' exceed the height of the Arc de Triomphe. See Heffernan (1995).
10 This was the first time some these national symbols were used on public buildings.
11 See for instance Sheftall (2010).
12 The idea that preserving the devastated landscape would be the 'best memorial of all' was widespread, especially amongst those responsible for the Dominion memorials. It was expressed, poetically, as early as 1917 by Lutyens. See Worpole (2003: 166). In the end, some disturbed ground, as well as preserved or recreated trenches, were included at Delville Wood, Vimy Ridge and Beaumont-Hamel.
13 A concept proposed by French historian Pierre Nora, a *lieu de mémoire* is a site, ritual or artefact of national memory where the past is evoked and represented.
14 This process began with the 1931 Statute of Westminster, which gave legislative independence to Australia, Canada, the Irish Free State, New Zealand, South Africa and Newfoundland. Broadly speaking, although it had different valency, popular interest in the war in Canada and Australia has mirrored that in Britain, whereas in South Africa it has largely disappeared.
15 The quality of bricks has been a recurring problem; those at Thiepval have been replaced twice in the memorial's lifetime, those at Villers-Bretonneux once. Extensive reconstruction of Vimy Ridge memorial's foundations was required in the last decade.
16 For instance, the recent rerouting of a department road to improve visitor access to Villers-Brettoneux. On this topic, see Gough (2007).
17 The exception to this is the South African monument at Delville Wood, where a new museum was constructed in the 1980s. According to the memorial's website, this was developed because the WGC felt South Africa was not keeping pace with the interpretative efforts at other ex-Dominions memorials.
18 See for instance Fathi (2014).
19 Including Britain itself. An elegant new visitor center was inaugurated at Thiepval in 2004. See www.thiepval.org.uk; accessed July 2015.

20 The black combatants were until now commemorated at Arques-le-Bataille in Normandy.
21 See for instance Scates (2006), Seaton (2000), Sumartojo (2014) and Winter (2006, 2012). Systematic records of visitor numbers and profiles at the large memorials are remarkably sparse and uneven.
22 Sources as for note 7. Although it varies by month, on average, recently more than 50 per cent of Vimy Ridge's visitors have arrived in groups.
23 That is, "our relations to sky, fire and earth, [and] the myriad ways of . . . relating materials and structure to all those elements and establishing systems of order (including disorder)".
24 Winter (2012).
25 Because the pastoral imagination represents a desire for a sense of belonging, it is a significant means by which we relate to the past (Heyde 2015: 190).
26 See, for instance, Jones and Cloke (2002).
27 This ambiguity between landscape as a material object and landscape as cultural projection has always been important in the battlefield memorials. While the evocation of the peaceful English landscape at the battlefield memorials helped comfort the bereaved, it also undermined popular anger against the political order responsible for the death of those remembered.
28 Reinforced by proscription of any commercial activity in the original deeds transferring these sites to the nations in question.
29 This co-presence of others creates a powerful emotional tension in the contemporary visitor, exemplifying Maurice Halbwach's seminal observation that collective memory is most powerful when it originates in joint action.
30 For example, "I came across some lad, an only son like ours killed so far from home". (Scates 2006: 111).
31 For example, "the waste of young lives . . . the sheer volume of death . . . lives in me still". (Scates 2006: 111).
32 For a comprehensive discussion of the social practices and performances involved in these groups' visits, see Iles (2006).
33 On the rare occasion when this principle is violated, as in the recent replanting of Villers-Brettoneux's allées, considerable public outcry ensues. See www.theaustralian.com.au/national-affairs/defence/villers-bretonneux-plots-stripped-bare/story-e6frg8yo-1225809983785. On the 'more-than-representational' affects associated with ongoing reconstitution of sites, see Edensor (2010, 2011).
34 See for instance Castree (2014).
35 As he observes, restored landscape feature such as ponds and forests provide a direct link to war poems and memoirs.
36 It is important to recognize how much this is a product of changing structures of seeing brought about by increasing urbanization. Once quite common, this kind of *uninterrupted relationship between third and second nature*, is no longer part of the life experience of most who visit.
37 Conventional mowers cannot cope with the undulating terrain, and could trigger unexploded, still buried, ordinance.
38 For instance interactive cell phone applications visitors can download before arriving at the memorial.
39 During the War, some colonial soldiers developed an affection for countries like France precisely because they were the antithesis of Britain. (White 1986: 44).

Bibliography

Anderson, S. (1995) 'Memory in architecture', *Daidalos*, 58: 23–37.
Berrizbeitia, A. and Pollak, L. (1999) *Inside/Outside: Between Architecture and Landscape*, Gloucester, MA: Greenwood Press.

Cadot, C. (2014) 'War afterwards: The repression of the Great War in European collective memory', in S. Sumartojo and B. Wellings (eds.), *Nation, Memory and Great War Commemoration: Mobilizing the Past in Europe, Australia and New Zealand*, Bern: Peter Lang, 259–272.

Castree, N. (2014) *Making Sense of Nature*, Abingdon: Routledge.

Chakrabarty, D. (2007) *Provincializing Europe: Postcolonial Thought and Historical Difference*, New York: Princeton University Press.

Conan, M. (2003) 'Garden and landscape design: From emotion to construction of self', in M. Conan (ed.) *Landscape and the Experience of Motion*, Washington, DC: Dumbarton Oaks, 1–34.

Corner, J. (1999) 'Paradoxical measures: The American landscape', *Architectural Design Profile*, 124: 46–49.

Della Dora, V. (2008) 'Mountains and memory: Embodied visions of ancient peaks in the 19th Century', *Transactions of Institute of British Geographers*, N.S. 33: 217–232.

Edensor, T. (2005) 'The ghosts of industrial ruins: Ordering and disordering memory in excessive space', *Environment & Planning D: Society and Space*, 23: 829–849.

Edensor, T. (2010) 'Networks of materiality; circulating matter and the ongoing constitution of the city', in M. Guggenheim and O. Söderström (eds.), *Reshaping Cities: How Global Mobility Transforms Architecture and Urban Form*, Abingdon: Routledge, 211–230.

Edensor, T. (2011) 'Entangled agencies, material networks and repair in a building assemblage: The mutable stone of St Ann's Church, Manchester', *Transactions of the Institute of British Geographers*, N.S. 36(2): 238–252.

Fathi, R (2014) '"A piece of Australia in France": Australian authorities and the commemoration of Anzac day at Villers-Brettoneux in the last decade', in S. Sumartojo and B. Wellings (eds.), *Nation, Memory and Great War Commemoration: Mobilizing the Past in Europe, Australia and New Zealand*, Bern: Peter Lang, 273–290.

Foster, J. (2004) 'Creating a temenos, positing South Africanism: Material memory, landscape practice and the circulation of identity at Delville Wood', *Cultural Geographies*, 11: 259–290.

Geurst, J. (2010) *Cemeteries of the Great War by Sir Edwin Lutyens*, Rotterdam: 010 Publishers.

Gibson F. and Kingsley Ward, G. (1989) *Courage Remembered: The Story behind the Construction & Maintenance of the Commonwealth's Military Cemeteries & Memorials of the Wars of 1914–1918 and 1939–1945*, London: HMSO/McClelland & Stewart.

Gough, P. (1998) 'Memorial landscapes as dramaturgical spaces', *International Journal of Heritage Studies*, 3(4): 199–214.

Gough, P. (2007) 'Contested memories, contested sites: Newfoundland and its unique heritage on the Western Front', *The Round Table*, 393: 693–705.

Harootunian, H. (2007) 'Remembering the historical present', *Critical Inquiry*, 33(1): 471–794.

Heffernan, M. (1995) 'For ever England: The Western Front and the politics of remembrance in Britain', *Ecumene*, 2: 293–323.

Heyde, S. (2015) 'History as source of innovation in landscape architecture: The First World War landscape in Flanders', *Studies in the History of Gardens & Designed Landscapes*, 35(3): 183–197.

Hunter, V. (2015) 'Worlding the beach: Revealing connections through phenomenological movement inquiry', in C. Berberich, N. Campbell and R. Hudson (eds.), *Affective Landscapes in Literature, Art and Everyday Life: Memory, Place and the Senses*, Farnham: Ashgate, 189–206.

Iles, J. (2006) 'Recalling the ghosts of war: Performing tourism on the battlefields of the Western Front', *Text and Performance Quarterly*, 26(2): 162–180.

Ingold, T. (1993) 'The temporality of landscape', *World Archaeology*, 25(2): 152–174.

Jameson, J. (1991) *Postmodernism, or, the Cultural Logic of Late Capitalism*, Durham, NC: Duke University Press.

Johnson, D. (1919) 'A geographer at the front', *Natural History*, 19(6): 511–521.

Jones, O. and Cloke, P. (2002) 'The non-human agency of trees', in *Tree Cultures: The Place of Trees and Trees in Their Place*, Oxford: Berg, 47–71.

Keegan, J. (1997) 'England is a garden', *American Scholar*, 66(3): 335–348.

Lefebvre, H. (2004) *Rhythmanalysis: Space, Time and Everyday Life*, translated by S. Elden and G. Moore, London: Continuum.

Lipstadt, H. (1999) 'Learning from Lutyens: Thiepval in the age of the anti-monument', *Harvard Design Magazine*, 10: 65–70.

Miller, K. (2012) *Almost Home: The Public Landscapes of Gertrude Jekyll*, Charlottesville: University of Virginia Press.

Morris, M. (1997) 'Gardens "For Ever England": Landscape, identity and the First World War British cemeteries on the Western Front', *Ecumene*, 4(4): 410–434.

Mycock, A., Sumartojo, S. and Wellings, B. (2014) '"The centenary to end all centenaries": The Great War, nation and commemoration', in S. Sumartojo and B. Wellings (eds.), *Nation, Memory and Great War Commemoration: Mobilizing the Past in Europe, Australia and New Zealand*, Bern: Peter Lang, 1–24.

O'Shea, S. (1997) *Back to the Front: An Accidental Historian Walks the Trenches of World War I*, New York: Walker & Co.

Park, O., Davidson, T. and Shields, R. (2011) 'Introduction', in O. Park, T. Davidson & R. Shields (eds.) *Ecologies of Affect: Placing Nostalgia, Desire and Hope*, Waterloo: Wilfrid Laurier University Press, 1–15.

Riegl, A. (2004) 'The modern cult of monuments: Its character and origins (1928)', in V. Schwartz and J. Przyblyski (eds.), *The Nineteenth-Century Visual Culture Reader*, New York: Routledge, 56–59.

Scates, B. (2006) *Return to Gallipoli: Walking the Battlefields of the Great War*, Melbourne: Cambridge University Press.

Seaton, A.V. (2000) '"Another weekend away looking for dead bodies": Battlefield tourism on the Somme and in Flanders', *Tourism Recreation Research*, 25: 63–77.

Sheftall, M.D. (2010) *Altered Memories of the Great War: Divergent Narratives of Britain, Australia, New Zealand & Canada*, London: I.B. Tauris.

Shields, R. (2011) 'The tourist affect: Escape and syncresis on the Las Vegas strip', in O. Park, T. Davidson and R. Shields (eds.), *Ecologies of Affect: Placing Nostalgia, Desire and Hope*, Waterloo: Wilfrid Laurier University Press, 105–126.

Sumartojo, S. (2014) 'Anzac kinship and national identity on the Australian remembrance trail', in S. Sumartojo and B. Wellings (eds.), *Nation, Memory and Great War Commemoration: Mobilizing the Past in Europe, Australia and New Zealand*, Bern: Peter Lang, 291–306.

Turner, V. (1974) *Dramas, Fields and Metaphors: Symbolic Action in Human Society*, Ithaca, NY: Cornell University Press, 166–230.

Wallis, J. (2015) 'Great-grandfather, what did you do in the Great War?: The phenomenon of conducting First World War family history research', in B. Ziino (ed.), *Remembering the First World War*, Oxford and New York: Routledge, 21–38.

Whatmore, S. and Hinchliffe, S. (2010) 'Ecological Landscapes', in D. Hicks & S. Beaudry (eds.), *The Oxford Handbook of Material Culture Studies,* Oxford: Oxford University Press, 440–458.

White, R. (1986) 'Bluebells and fogtown: Australian's first impressions of England, 1860–1940', *Journal of Australian Cultural History,* 5: 44–59.

Winter, C. (2006) 'Battlefield visitor motivations: Explorations in the great war town of Ieper, Belgium', *International Journal of Tourism Research,* 13: 164–176.

Winter, C. (2012) 'Commemoration of the Great War on the Somme: Exploring personal connections', *Journal of Tourism and Cultural Change,* 10(3): 248–263.

Worpole, K. (2003) *Last Landscapes: The Architecture of the Cemetery in the West,* London: Reaktion.

Wylie, J. (2007) 'The spectral geographies of W. G. Sebald', *Cultural Geographies,* 14(2): 171–188.

10 Local complications

Anzac commemoration, education and tourism at Melbourne's Shrine of Remembrance

Shanti Sumartojo

Introduction

The year 2015 was extraordinary for Australia's Anzac narrative. The centennial anniversary of the Australia and New Zealand Army Corps (Anzac) landing at Gallipoli, the 'Anzac Centenary', saw an estimated $500 million spent on commemorative activities and memorial sites around Australia and the world, including the announcement of a new $100 million museum at the main Australian National Memorial on the Western Front in Villers-Bretonneux, France (Honest History 2015). The anniversary permeated deeply into many aspects of everyday Australian culture, with new television dramas, special edition currency, museum exhibitions, artworks and performances, and publications of all kinds efflorescing into the public realm. Anzac iconography was ubiquitous, with even the Melbourne zoo inviting visitors to 'celebrate Anzac Day with some of Australia's iconic wildlife' accompanied by a photo of a wombat, who appeared to be smiling.

This proliferation of Anzac references did not go uncritiqued. Commentary by journalists and academics condemned widespread 'Anzackery' (Daley 2015) and the 2015 'parody of remembrance' (Scates 2015) that deflected public attention from the traumatic realities of war and other serious issues of national identity, such as reconciliation between indigenous and non-indigenous people (see Wellings and Sumartojo 2017). Further criticism came in the book *Anzac's Long Shadow* (Brown 2014), which argued that high government and private sector funding on Anzac commemoration diverted badly needed resources away from younger veterans of more recent conflicts suffering with disability and illness as a result of their military service.

Nevertheless, Anzac Day, 25 April, saw a peak of commemorative events, with dawn services and mid-morning marches around the country and worldwide. Blanket television coverage or events at sites such as Melbourne's main war memorial, the Shrine of Remembrance, reached a much wider audience than the approximately 85,000, myself included, who attended in person. The Shrine was one of several well-known sites (including memorials in Sydney, France and of course Gallipoli) that acted as a symbolic and material backdrop to collective commemorative rituals with national and international reach. In this chapter, I take up the case of Melbourne's Shrine of Remembrance, discussing its role as a commemorative, pedagogical and tourist site. I argue that these multiple purposes create a set of unavoidable 'local complications' that unsettle the traditional meaning of the site, but that also gesture towards its possible future as it moves forward from the First World War Centenary.

Anzac nationalism and the Shrine of Remembrance

The history, politics and social impact of Anzac has been subject to sustained and intense scholarship, particularly during the Great War centenary period that began in August 2014 (Scates 2006; Holbrook 2014; Brown 2014; Sumartojo and Wellings 2014; Australian Historical Review 2015). This has included research on the spatial aspects of Anzac commemoration and its relationship with Australian identity, including the history of Anzac memorials (Inglis 1998; Scates 2009) and their architectural and landscape settings (Moriarty 2009; Stephens 2012), the significance and symbolism of commemorative sites for visitors (Winter 2009; Sumartojo 2014), non-representational and affective aspects of Anzac sites (Waterton and Dittmer 2014; Sumartojo 2015), and the relationship between family history and Australian identity (Holbrook and Ziino 2015). This body of work recognises the symbolic importance of Anzac memorials, as well as their practical value for family members of former service-people, particularly those grieving for the military dead.

The Anzac narrative is anchored in the First World War, a conflict that was simultaneously intimate and distant for Australia. Although family grief was powerful and widespread, only two bodies of Australian war dead were ever repatriated: General Bridges, who was killed at Gallipoli in 1915, and the body of an unknown soldier, exhumed from Adelaide cemetery on the Western Front and reinterred in the Australian War Memorial in Canberra in 1993. This meant that for most Australians, the loss of family members in the First World War was expressed as a 'distant grief' (Ziino 2007), manifest and recognised in local memorials that came to represent foreign gravesites. The story of family loss, of ordinary men and women caught up in the conflict, was central in then Prime Minister Paul Keating's speech at the burial of an unknown soldier from this war: 'real nobility and grandeur belong not to empires and nations but to the people on whom they, in the last resort, always depend' (Keating 1993). As Inglis (1998: 41–42) remarks, First World War memorials in Australia 'had to serve both as a statement about the soldiers and as a substitute for their graves – a cenotaph, an empty tomb'. Anzac memorials across the country served this function, with most small towns erecting some type of memorial statue, arch, tree-lined avenues, or honour roll in local churches, schools, businesses and government buildings (Inglis 1998). In the big cities, such sites were centrally located and hosted major annual ceremonies.

Although Canberra had been selected as the site for the new Australian capital in 1908, as the First World War ended Melbourne was still the seat of the Australian government and many believed that a national memorial should be built there.[1]

In 1922, a competition was held and a design chosen (Scates 2009). The Shrine was completed in 1934, on a hilltop in the Domain, a large park of Crown Land south of Melbourne's central district and the Yarra River, and a visual southward extension of the line of the city's axial Swanston Street. The Shrine's location just out of the city centre was intended to create a physical and symbolic distance to mark it out as special (see Figure 10.1). To counter criticisms that it should have a more central location, one of the designers pointed out the symbolic importance of the physical approach to the site: 'The very fact of its aloofness, and the fact that something of a pilgrimage is necessary to approach it, will prevent the possibility of the memorial [ever] being taken for granted' (Hudson cited in Scates 2009: 66).

This slight remove from the bustle of the city, the lawns and plazas that surround it, and the protection in planning legislation of the space around the Shrine and vistas towards and away from it, all work to emphasise its significance visually, spatially and symbolically. Benjamin (2003) describes the site as 'a secular temenos', a contemporary reworking of the Classical relationship between the temple and its surroundings that, in the case of the Shrine, uses landscape architecture to signal a transition onto 'sacred' space. This delineation between the quotidian and reverential is an important aspect of many memorials located in cities, and for the Shrine this occurs through landscape elements including grassy lawns, paved walkways for circulation and the removal and addition of major trees.

Figure 10.1 The view from the Shrine looking north to the centre of Melbourne along the main city axis of Swanston Street.

Figure 10.2 The front of the Shrine of Remembrance.

The Shrine's built environment acts as a dramatic backdrop to a range of commemorative activities. The main approach from St Kilda Road is framed by tall conifers, which funnel the visitor's gaze forward and up to the stepped, pyramidal shape of the roof. The inclined, tree-lined approach opens into a wide, symmetrical plaza flanked by pillars and with an Eternal Flame to one side. The eye is constantly pulled upwards by the site's elevation, its rows of shallow steps and regular Doric columns crowned by a heavily carved classical tympanum (see Figure 10.2). On the north, city-facing side, this depicts an allegorical group of a female national mother figure flanked by an older shepherd anxious for the safety of his flock and a young man keen on the excitement of war (Shrine 2015a). The symbolic nature of the building is signaled directly in its form and decoration (Sumartojo and Stevens 2016).

A recent major refurbishment and extension to the Shrine, which re-opened on Remembrance Day 2014, cost $45 million. Designed by architects Ashton Raggatt McDougall and landscape architects Rush Wright and Associates, the redevelopment of the undercroft space added a new Galleries of Remembrance (a museum space in the reclaimed undercroft of the building) that won several design and architecture awards.[2] It also saw visitor numbers increase by 15 per cent. The transformed visitor facilities and exhibition space are now accessed through a submerged courtyard, one of four symmetrical sunken spaces at each corner of the building, that symbolise different aspects of the Anzac narrative, including a massive metal poppy that appears to float over the school groups entrance (see Figure 10.3).

Figure 10.3 The visitors' entrance with its umbrella-like poppy sculpture.

The Shrine, however, is not designed solely as a memorial. Its location and urban context also means that it is a prominent tourist attraction, and in 2014–15, it received around 850,000 visitors (Shrine 2014–15). Recent changes have seen it embrace a new identity as a tourist destination, although this role does not always sit easily with its simultaneous function as a commemorative site for veterans and families of service-people. Thus, commemoration at the Shrine of Remembrance is complicated by its multiple functions, and how it manages this will shape how Australian and state of Victoria war service continues to be remembered. The following sections focus on the Shrine of Remembrance's unique spatial and symbolic character, and how this has helped shape how the 'Anzac story' is understood and experienced at commemorative events, drawing my own first-hand experience, as well as media coverage. I will also discuss the Shrine's role as an education and tourism destination, drawing out how this complicates an apparently straightforward and long-standing role as a national commemorative site. I conclude with some observations on the strategies that the Shrine uses to attempt to reconcile these, and what this suggests for its role beyond the First World War Centenary.

Commemoration at the Shrine

In addition to many smaller ceremonies, there are two main commemorative events at the Shrine every year: 25 April, Anzac Day, and 11 November, Remembrance

Day. The former is by far the most popular, with 60,000 attendees in 2014 and over 85,000 in 2015. The 2015 ceremony marked the centenary of the landing at Gallipoli, held to be the key generative moment of contemporary Australian nationalism. Seal (2011: 50–51) recounts the different explanations for its genesis, linked to military traditions of the dawn muster of troops, or the very early attack on Gallipoli by Anzac troops. He argues that the Service: 'conflates the sacred and the secular, the military and the civilian with the official and the folkloric in an especially charged moment of time that involves significant numbers of people throughout the country and beyond'. I have written elsewhere about the affective impact of the Service's rhythms of speeches, collective repetition of ritual texts, music and silence combined with built environment and experiential elements such as weather, light and the presence of a large crowd (Sumartojo 2015; Sumartojo and Stevens 2016). In this work, I argued that commemoration should be understood as linking meaning to the sensory experience of participating individuals, granting this equal importance with official narrative through events and environments intended to work on our senses and feelings in purposeful ways.

At the Shrine Dawn Service in 2015, I stood amongst the crowd clustered in front of a big screen that showed the façade of the Shrine building and the speakers and musicians arrayed on it. I could not see the building itself, but could hear clearly above the rustling of people's umbrellas and rain ponchos. The main address, by Vietnam veteran Neville Clark, invoked the Gallipoli dawn assault, describing the experience of nervous soldiers waiting for orders to attack:

> The moon is down, it is 3am. They know they will have an hour of darkness before dawn will silhouette the jagged darkness of Sari Bair . . . a captain of the Turkish forces is straining his eyes seaward.
>
> (transcribed from video, ABC News 2015)

Clark's theatrical delivery from the podium on the steps of the Shrine, and also played on big screens dotted around the site, set a tone of immediacy that sought to directly connect participants to the soldiers at Gallipoli through a dramatic present-tense act of story-telling. Winter (2010: 15) argues that 'the performative act of remembrance is an essential way in which collective identities are formed and reiterated', and Clark's performance enrolled attendees in an *experience* of commemoration that was as much affective as it was narrative (see also Wellington 2016). This affective intensity is one of the main features of the Dawn Service and its combination of bodily, sensory and cognitive engagement with Anzac ritual, that encourages participants to find personal meaning in the event. The metaphoric and mnemonic potential of such spatial experiences is evident, for example, in the popular reactions that draw together physical and affective experiences, as in the repeated descriptions of Anzac Day as 'moving' (see McKenna and Ward 2007).

As Marshall (2004: 38) remarks, 'we use our senses to forge connections with our physical environment and develop our sense of place . . . Remembrance, because it is experienced through the senses, is one such embodied state'.

However, this state does not always cleave to proscribed ways of experiencing Anzac commemoration. As I stood in the dripping rain at the Shrine in 2015, shivering gently, I envied my companion, who had found a dry spot under a tree. I jumped at the sound of rifle shots that punctuated the singing of *Abide With Me* and wondered at the appropriateness of this addition to the well-known hymn. Later, searching for a coffee in a chilly but crowded post-ceremony central Melbourne, I grumbled about the disconnect between a 'major' event and the opening times of nearby cafés. All these embodied experiences worked to complicate the official narrative of reflection on service and sacrifice that Anzac Day is designed to engender by distracting me from the official purpose of the event. My sensory perceptions did not perfectly align with the representational aspects of the Dawn Service, and although I knew that it was a solemn occasion for collective remembrance, my chilly, damp feet and the tensing of my body at the startling sound of rifle shots pulled against the somber, reflective atmosphere.

By contrast, Remembrance Day 2015 was a much less spectacular and atmospheric event. There were only about 5,000 attendees at the mid-morning ceremony, many of whom were school students. In her brief address, the State of Victoria Lieutenant-Governor Marilyn Warren said that 'it was a magnificent thing . . . to see so many young people at the ceremony, for the remembering now was in their hands' (in Wright 2015). The students in the rear seats near where I was standing were still, but not attending closely to the ceremony. The dark suits of political and bureaucratic attendees on the steps of the Shrine formed a backdrop to the speaker's podium, and young Army cadets struggled to stand to attention for the duration of the ceremony. The printed program laid out the 'order of service', but it felt less scripted than the Dawn Service, which had more familiar rhythms and conventions. The smaller crowd, the predominantly student participation and the less settled routines of the event combined to suggest that Remembrance Day is a much less important event in the annual commemorative calendar than Anzac Day.

Somewhat paradoxically, the Shrine's popularity for major commemorative events is also identified as a risk by its management, who are concerned about a 'lack of public awareness and understanding of the role of the Shrine including a focus on key commemorative days . . . instead of year round commemoration' (Shrine 2013–2014: 8). This helps to explain the regular 'Ray of Light' ceremonies inside the main Shrine building that have always been a feature of the design of the central chamber. At these events, performed at half-hour intervals for visitors and school groups, a volunteer guide begins by explaining the significance of 11 November. The building is designed so that a beam of sunlight falls on the word 'love', part of the inscription 'greater love hath no man', set into the floor of the Shrine's inner sanctum. This occurs at 11.00 am on 11 November every year, but to demonstrate the symbolism on demand, the ceremony is scheduled to repeat regularly with an electric light simulating the fall of sunlight onto the word 'love'. This is accompanied by a recording of a bugle playing the Last Post.

On my visit, a group of students stood with heads bowed, listening to the recorded explanation and music, and dutifully observing the fall of the spotlight onto the engraved word 'love'. The sound echoed around the central chamber, which stretched up in a pyramid shape over our heads, lined round with friezes depicting the different military services at work. Although it felt somewhat contrived to be listening to a recording and looking at a pre-programed spotlight, the visitors stood silently, and the sound and space of the room worked to emphasise the sensory effect. The Shrine's built environment and its sensory affordances of echoing bugle music and shining light co-mingled narrative and affect to both commemorative and pedagogical ends. Compared to the Dawn Service, in this briefer commemorative ceremony the physical setting rather than the historical resonance of the date generates the main affective charge. There is similar use of repeated text and music and a shared setting at a memorial site, although this ritual takes place in a highly decorated interior space. All the ceremonies at the Shrine thus share some similar sensory elements, although the different scale of participation and larger cultural significance of time and date make them distinctly different Anzac experiences.

Taken together, the building of the Shrine, its immediate landscape setting and its wider urban spatial context, along with its commemorative activities, give a straightforward message of the importance of Anzac in contemporary Australia, and the accepted contours of the narrative. Even within the variation of Anzac and Remembrance Days, and the repeated and choreographed 'Ray of Light' ceremony, as a visitor I was struck that these events were a central intended function of the spaces of the Shrine. This is reflected and reinforced in its architecture, decoration, setting, location and formal use as a physical and aesthetic setting for commemorative events.

However, the Shrine's role is more complex than simply one of war remembrance, and these other functions – particularly for education and tourism – can pull against its stated central mission to serve veterans and their families. As I will discuss in the next two sections, reconciling the Shrine's various roles and uses is an ongoing challenge for its leadership and one that shapes how Anzac is both constructed and remembered in contemporary Melbourne.

School students and affective Anzac heritage

When I attended the 2015 Remembrance Day Service, one function of the Shrine was immediately evident: its role in educating school students about Australian military history (see Figure 10.4). I recorded the students' attendance in my notes:

> The seats nearest me were full of high school students in bright blue jackets. They sat still but not particularly attentive, slouched, sometimes whispering to each other. The majority of people there were school groups, although the seats near the front were a mix of un-uniformed bodies. Everyone was positioned to look up at the Shrine, which was sharp against the cloudy grey sky.

This involvement is seen to have both a pedagogical function and one of renewing interest in war commemoration and military history in general, rather

Figure 10.4 Rows of students in school uniforms look up at the Shrine on Remembrance
    Day 2015.

than being limited solely to First World War remembrance, addressing perpetual
concerns on the part of the Shrine that its symbolic function will atrophy as generations of veterans die. As a result, education is a growing aspect of the Shrine's
duties, and is reflected in the legislation that sets out the responsibilities of the
Shrine's trustees, who have three main roles: to manage and care for the Shrine
itself; to organise commemorative activities; and to develop and implement public
education programs. Accordingly, in 2013–14, the over 51,000 Victorian school
students experienced the Shrine's education program. This program links student
learning to particular conflicts, and is unabashed about its goals in doing so:

> With a continued emphasis on commemorating the service and sacrifice of
> Australian service men and women, students gain valuable insights into Aus-
> tralian history and changing national identity and to reflect on the personal
> qualities exemplified through the Spirit of Anzac.
>
> (Annual Report 2013–14)

The conflation of the national curriculum with the inculcation of particular ways
of understanding the Australian experience of war – through the lens of 'service
and sacrifice' rather than, for example, violence or venereal disease – has of
course been criticised in Australia. Most notably, Lake's (2010) *What's Wrong
with Anzac*, argued that the 'Anzac myth' had militarised Australian history to

the detriment of other important national narratives. The history of the Anzacs is commonly taught to both primary and secondary school students, and the Shrine has an outreach program that targets its education programs at rural and disadvantaged schools that would find it difficult to bring students to Melbourne for a visit. However, amongst others, McKenna (2014) has argued that this expenditure of resources privileges official martial histories over others, occluding other important issues of national identity such as Aboriginal reconciliation.[3] Indeed, Lake (2010: 137) remarks that in Australia, 'schoolchildren are now conceptualised as the inheritors of the Anzac spirit and its custodians', and identifies the crucial role of government in creating pedagogical materials and distributing them to schools nationwide since around 2000. The Shrine publishes free materials for teachers to use with students that align with the Victorian state curriculum, and that include activity sheets, essay questions and visual material for students. Teachers can book free guided tours at the Shrine, led by volunteer staff, and sessions where students can handle historical objects.

Strategies to engage school students often use material culture or first-hand narrative to personalise the stories of war service and engage students with Anzac narratives.[4] These in-house programs are tailored to appeal directly to schoolchildren's interests and are aimed at students of all ages. Children up to Year 4 (around 10 years old) can learn about the 'loyal friends', the animals that have served in war. Teachers of older primary school students can select the 'Spirit of Anzac' session that 'includes the handling of various items of equipment and uniforms to enable students to empathise and understand what life would have been like . . . 100 years ago' (Shrine 2015b). Within these educational narratives, inspiring stories about animals and material objects are used to pique children's interest and link them to a version of the national past that is told through personal stories and artefacts. Older students are offered tours of the 'Galleries of Remembrance', or 'hands on' experience of objects from the Second World War and Vietnam War – the latter program entitled 'Only 19' emphasizing the age-specific connection to students' own lives.

This focus on historical objects is a common approach for heritage organisations to create an emotional connection between contemporary visitors and the lives of the people whose stories they tell (see Wallis and Taylor, this volume). Indeed, at the Australian War Memorial, such affective strategies stretch to the multimedia displays that envelop visitors in sound, light and movement that link narrative to sensory perception, making visitors literally *feel* as if their experience physically echoes that of former service-people (see Waterton and Dittmer 2014). In Belgium, the In Flanders Fields Museum in Ypres uses interactive displays, film of actors in historical costumes speaking straight to visitors and tailored sensors to provide personalised versions of historical events that appear designed to collapse the temporal distance between present-day visitors and the historical objects and individuals presented. Indeed, Witcomb (2013: 256) has argued for the importance of the 'sensorial, embodied forms of knowledge that express themselves though feelings in response to the material, aesthetic, and spatial qualities of the exhibition/interpretation', showing how they 'play a role in the production of meaning' that moves beyond the 'more explicit rational, information based content of the display'. Such 'forms of knowledge' are created through the experience of handling objects, trying on clothing or helmets, reading intimate records such as diaries and letters, or studying candid photographs that capture service-people doing the work of war. The affective and sensory aspects of these material encounters are a way to make Anzac history 'stick' for participating students.

The large number of student visitors demonstrates the relationship between the Shrine's educational program and its commemorative mission. On the one hand, it speaks directly to an ongoing concern that an important user group of the site – young people – might be flagging in their interest in or knowledge of the Anzac narrative, and that this is an important lesson that requires constant reinforcement. As the ranks of veterans of twentieth century conflicts thin, commemorative organisations such as the Shrine have sought to engage young people directly through schools education programs, prize-giving schemes, or displays that highlight the young age of many First World War 'diggers', an Australian term for the enlisted men who served in this and other military conflicts. The Shrine appoints Year 9 students (who are around 15 years old) to be 'Young Ambassadors' to 'encourage their peers and other young people in their communities to gain a greater understanding of the importance of remembering and honouring Australia's service community and military history' (Shrine 2015c). All these activities link pedagogy and commemoration through the Shrine's distinctive memorial precinct.

In the activities I have described here, pedagogy, youth and Anzac remembrance are tightly intermeshed to advance the Shrine's commemorative remit. However, the third main function of the Shrine is much less straightforward in its relationship with commemoration. Tourism at the Shrine is a growing but complicated aspect of what happens there, and one that presents new challenges to its commemorative role.

Tourism

In addition to commemoration and education, tourism is the third main function of the Shrine. As Winter (2009) points out, people do not just visit war memorials to remember the dead or learn more about military history; tourism is another common reason people visit commemorative sites, and a large draw for the Shrine's many visitors. Tourism at commemorative sites can be fraught with concerns about trivialisation or commodification of war sites (Henderson 2000), or embarrassment or unease about wartime activities (Cooper 2005). It can also be stage-managed to 'mobilise affect in a ritualised affirmation of the national identity' (Hudson 2009). Many tourist visits are to historical sites of war such as former battlefields, concentration camps or other locations of conflict-related civilian trauma. For Anzac commemoration, tourism to the Gallipoli Peninsula and sites such as Villers-Bretonneux, Fromelles and Pozières on the Western Front is an important part of local economies and infrastructure as much as a manifestation of personal mourning or national remembrance. However, the Shrine of Remembrance differs from these in its location in a major city. The addition of a visitors' centre to the site in 2003 was the beginning of a process that culminated in 2014 with improved visitor facilities and the refurbishment and development of the display space of the new Galleries of Remembrance. Its popularity is enhanced because it can readily be incorporated into a bigger tour of the city, given its easy walking distance from the city centre and good road access, and it is near other prominent tourist sites, including the National Gallery of Victoria and the Botanical Gardens.

There appears to be an awkward relationship between the goals of the Shrine to commemorate and educate on the one hand and its status as a popular tourist site on the other. At the Shrine, some people who visit have a personal connection to Anzac, such as a relative who served in the military. However, in one of the few studies of visitors to the Shrine of Remembrance, Winter (2009: 561) found that while the emotional experience may be intense, the pedagogical goals of informing visitors about the Australian war experience are less successful, with the possibility that 'some visitors may have confused factual information, and perhaps conflated the meanings of the First and Second World Wars'. Many people had little knowledge of Australian military history before their visit and their feelings about the various conflicts in which Australia has fought stem, at least in part, from the conventional ways in which these wars are portrayed in popular culture.

The figures that personify the tension between commemoration and tourism are those of the volunteer guides who explain and interpret the displays and 'Ray of Light' ceremony, give twice-daily guided tours, and deliver other aspects of the Shrine's schools and public education programs. The demands of tourism, however, place volunteer guides into the position of not only presenting narrative, but also of policing behavior and enforcing codes of conduct. These expectations are outlined in this text from Shrine's (2015) website:

> The Shrine welcomes all visitors to enjoy their visit to the memorial and respects the cultural diversity of our visitors. We do ask that the behaviour of visitors whilst at the Shrine, remains respectful to the nature of the Shrine as a sacred place for commemoration.

This statement hints at the diversity of visitors to the site, but also some of the challenges that they present. The Shrine's mission to commemorate, and by extension serve the needs of veterans and their families to do so in an 'appropriate' setting, can be disrupted by large groups that that have little foreknowledge of the history the Shrine represents or the unwritten cultural expectations of behavior there. The high numbers of visitors can exacerbate this. Even domestic school students, usually already acculturated into the 'Anzac legend' and with teachers monitoring their behavior, can disturb the expectations of volunteers with adolescent conversation and movement. Another group that could potentially disrupt the Shrine's affective aims is foreign visitors. Research on Chinese tourism in Victoria found that while the Shrine was on many itineraries, an average of only about 20 minutes was spent there on group visits (Weiler and Yu 2006). Although this research was done before the opening of the new Galleries of Remembrance, the itinerary-driven tourism experience of these tourists suggest they have little time to absorb the museum displays or immerse themselves in the 'sacred' qualities of the site.

The sheer numbers and the range of visitors has been a challenge, albeit probably a welcome one, for the Shrine's management. High visitor attendance affirms the popularity and relevance of the site, but also places demands on volunteers and staff. One response has been the development of a 'Galleries of Remembrance Highlights Tour' smartphone app available for download on the Shrine's website that can guide visitors through the site and highlight important displays in the Galleries. There has also been recognition by management that more volunteer guides are needed to help deliver educational programs and other visitor services. These efforts show how the Shrine is seeking to recognise, manage and provide for the diverse visitors it attracts.

Conclusion: local complications

Despite the monolithic design and clear and well-established Anzac narrative, local complications creep into and disrupt the spatial, narrative and affective logics of the Shrine of Remembrance. In this chapter I have discussed how the site's three roles – commemoration, education and tourism – sit alongside each other, how the Shrine accommodates their different demands, and some of the effects they may have on each other. Increasing visitor numbers and official requests that people respect 'the nature of the Shrine' (without actually specifying precisely what this is) suggest unease that these newer functions, which continue to grow in reach and audience, could disrupt the traditional and long-standing functions of the Shrine to commemorate Australian war service.

One way of thinking about how these functions fit together, and how they might simultaneously strengthen and disrupt each other, is by focusing on the affective aspects of the Shrine and the activities that occur there. Indeed, affect is an established frame through which commemorative sites can be understood (Todman 2009; Waterton and Dittmer 2014; Sumartojo 2015; Wellington 2016). Particularly during commemorative ceremonies, the sites of war remembrance contribute to particular sensory perceptions, historical narratives, rhythms of movement and physical locations to coalesce particular ways of feeling about national identity. However, in the case of the Shrine, the commemoration that the site hosts is complicated by the experiences of visitors that do not conform to the official versions; in my own account, for example, the dripping rain, cold feet and early morning sleepiness were all banal, sensory experiences that pulled against the official solemnity of the Dawn Service. As personal mourning for killed family members becomes less common, and as war veterans diminish in numbers and public visibility, commemorative ceremonies help to link the past to contemporary national life. However, the affective intensities that these ceremonies rely on for their impact also make the experiences individual, contingent, emergent and therefore unpredictable (Anderson 2014), complicating the narrative thrust of commemoration.

Additionally, the presence of large numbers of school students at the Remembrance Day ceremony is a reminder of the educational mission of the Shrine, and its close connection with commemoration. Through the figure of youth, in its official, school-uniformed version, education and commemoration augment each other. This helps to build new audiences for Anzac and reinforce national identity through a combination of various affective and narrative means: participation in ritual ceremonies at the Shrine; contact with objects, including intimate texts such as diaries and letters, intended to link contemporary individuals directly to people in the past; and pedagogical materials targeted at particular age groups that seek to demonstrate the abstract concepts of 'service and sacrifice'. While the Shrine regards its schools program as a central part of its mission, it also appears to recognises the potential disruption that students might present within the solemn atmosphere of the site, and so manages this by actively signaling the ways in which the students should experience the Shrine.

A less straightforward audience for the Shrine's message is made up of tourists, many of whom have no direct personal link to Australian war service, who may be from overseas or whose visits are very short. The major refurbishment and the addition of the Galleries of Remembrance both recognise this growing audience and seek to cater for it, even if this sits somewhat uneasily with the attempts by the Shrine to make a commonly understood affective experience central to its commemorative function. For example, the Shrine relies on its built and landscape environment to signal the 'appropriate' emotional responses in all visitors, that of reverence and reflection, even as they approach the site. In the Galleries, the display of personal objects; the use of sound, film and photographs; and innovations such as the Highlights Tour smartphone app work to encourage personal and emotional engagement alongside cognitive understanding. In this way, remembrance and education are blurred together with tourism at the Shrine, even if there is evidence that these multiple functions might sometimes pull against each other.

Scates (2009: 256) remarks that 'the meaning of the Shrine has never stayed the same. It is ironic perhaps that a building that so prized permanence should continue to change, challenge and evolve'. This chapter has suggested that the Shrine's continued relevance relies on this process of evolution, linked in part to the purposeful development of affective experiences for many types of visitors. In the First World War Centenary period, political will, and the funding that accompanies it, have so far been plentiful in helping the Shrine transform. The physical changes to the site have been accompanied by purposeful attempts to deploy affective strategies to enroll visitors in Anzac commemoration and build the relevance of the Shrine for future generations. However, as visitors to the Shrine continue to grow and potentially diversify, the memorial will have an ongoing challenge in the management of its complicated meanings.

Notes

1 Canberra was selected as the site of the Australian capital in 1908, and the building of a planned city began in earnest in 1913. The Australian Parliament moved to Canberra in 1927 and national Australian War Memorial was eventually built in 1941.
2 The Shrine redevelopment won five awards at the 2015 Victorian Architecture Awards, including the highest award offered by the Architecture Institute's Victoria Chapter. See

www.majorprojects.vic.gov.au/project/shrine-galleries-remembrance/. Accessed 4 May 2016.
3 One aspect of the tension between Anzac commemoration and Aboriginal reconciliation has been the almost non-existent recognition of Indigenous soldiers' service in the First World War, although this is slowly being addressed. See Wellings and Sumartojo (2017).
4 Scates (2006: 173) identifies a similar intimacy of students' engagement with battlefield tourism that contextualizes their own family histories of military service with the national past and mainstream interpretations of that past. Elsewhere, I have discussed how Anzac commemoration rests in large part on family remembrance, and the effect this can have on narrowing the use of Anzac by diverse Australians to connect to a shared national identity (Sumartojo 2014).

Bibliography

ABC News (2015) 'Anzac Day 2015: Rain fails to deter massive crowds at Melbourne commemoration', 25 April. Available from: www.abc.net.au/news/2015-04-25/record-crowd-expected-at-melbournes-anzac-day-dawn-service/6420488 (accessed 2 December 2015).

Anderson, B. (2014) *Encountering Affect: Capacities, Apparatuses, Conditions*, London: Routledge.

Australian Historical Studies (2015) Special issue on 'Remembering Australia's First World War' 46: 1–127.

Benjamin, A. (2003) 'A secular temenos', 1 September. Available at http://architectureau.com/articles/a-secular-temenos/ (accessed 14 May 2017).

Brown, J. (2014) *Anzac's Long Shadow: The Cost of Our National Obsession*, Melbourne: Black Inc.

Cooper, M. (2005) 'The Pacific War battlefields: Tourist attractions or war memorials?', *International Journal of Tourism Research*, 8: 213–222.

Hamann, C. (2015) 'Shrine of remembrance: Galleries of remembrance', *Architecture Australia*, 104(3): 24–31.

Henderson, J. (2000) 'War as a tourist attraction: The case of Vietnam', *International Journal of Tourism Research*, 2: 269–280.

Holbrook, C. (2014) *Anzac: The Unauthorised Biography*, Sydney: NewSouth Publishing.

Holbrook, C. and Ziino, B. (2015) 'Family history and the Great War in Australia', in B. Ziino (ed.), *Remembering the First World War*, London: Routledge, 39–55.

Honest History (2015) 'Budget 2015: Honest history factsheet: Centenary spending $551.8 million'. Available from: http://honesthistory.net.au/wp/budget-2015-honest-history-factsheet-centenary-spending-551-8-million/ (accessed 17 December 2015).

Hudson, C. (2009) 'Embodied spaces of nation: Performing the national trauma at Hellfire Pass', *Performance Paradigm*, 5(2): 1–27.

Inglis, K. (1998) *Sacred Places: War Memorials in the Australian Landscape*, Melbourne: University of Melbourne Press.

Keating, P (1993) *Remembrance Day 1993*. Available from: www.awm.gov.au/talks-speeches/keating-remembrance-day-1993/ (accessed 12 May 2016).

Lake, M. (2010) 'How do school children learn about Anzac', in M. Lake and H. Reynolds (eds.), *What's Wrong with Anzac? The Militarisation of Australian History*, Sydney: NewSouth Publishing, 133–160.

Marshall, D. (2004) 'Making sense of remembrance', *Social & Cultural Geography*, 5(1): 37–54.

McKenna, M. (2014) 'Keeping in step: The Anzac "resurgence" and "military heritage" in Australia and New Zealand', in S. Sumartojo and B. Wellings (eds.), *Nation, Memory and Great War Commemoration: Mobilizing the Past in Europe, Australia and New Zealand*, Bern: Peter Lang, 151–168.

McKenna, M. and Ward, S. (2007) '"It was really moving, mate": The Gallipoli pilgrimage and sentimental nationalism in Australia', *Australian Historical Studies*, 38(129): 141–151.

Moriarty, C. (2009) 'The returned soldiers' big': Making the Shrine of remembrance, Melbourne', in N. Saunders and P. Cornish (eds.), *Contested Objects: Material Memories of the Great War*, Abingdon: Routledge, 144–162.

Scates, B. (2006) *Return to Gallipoli: Walking the Battlefields of the Great War*, Cambridge: Cambridge University Press.

Scates, B. (2009) *A Place to Remember: A History of the Shrine of Remembrance*, Cambridge: Cambridge University Press.

Scates, B. (2015) 'Political rhetoric makes a parody of remembrance', 23 April. Available from: http://www.smh.com.au/comment/political-rhetoric-makes-a-parody-of-remembrance-20150421-1mqdc2.html (accessed 14 May 2017).

Seal, G. (2011) '". . .and in the morning. . .": adapting and adopting the Dawn Service', *Journal of Australian Studies* 35(1): 49–63.

Shrine of Remembrance (2015a) *Shrine Exterior*. Available from: www.shrine.org.au/The-Shrine-Story/Features-and-Memorials/Shrine-Exterior (accessed 16 November 2015).

Shrine of Remembrance (2015b) *Primary: Spirit of Anzac G5–6*. Available from: www.shrine.org.au/Education/Programs/Welcome-to-the-Shrine#sthash.FBr60zP1.dpuf (accessed 2 December 2015).

Shrine of Remembrance (2015c) *Young Ambassadors Program*. Available from: www.shrine.org.au/education/programs/young-ambassador-s-program#sthash.ikYL1Eln.dpuf (accessed 13 December 2015).

Shrine of Remembrance 2013–2014 Annual Report. Available from: www.shrine.org.au/Shrine/Files/06/06b4d92b-f2b2-4fcd-b40d-106157a51371.pdf (accessed 16 November 2015).

Shrine of Remembrance 2014–15 Annual Report. Available from: www.shrine.org.au/Shrine/Files/2d/2d5dd6b9-2e2d-48c5-97d4-52278bc93bc4.pdf (accessed 16 December 2015).

Shrine of Remembrance Act 1978 (Amended), Section 4(1AA). Available from: www.legislation.vic.gov.au/Domino/Web_Notes/LDMS/LTObject_Store/LTObjSt8.nsf/DDE300B846EED9C7CA257616000A3571/A0B815411CE8B95ACA257D0700038B15/$FILE/78-9167aa026%20authorised.pdf (accessed 18 November 2015).

Shrine of Remembrance Strategic Plan 2013–18. Available from: www.shrine.org.au/Shrine/Files/1d/1deca798-252a-4698-a866-71dbd73d6114.pdf (accessed 18 November 2015).

Stephens, J. (2012) 'Recent directions in war memorial design', *International Journal of the Humanities*, 9(6): 141–152.

Sumartojo, S. (2014) 'Anzac kinship and national identity on the Australian Remembrance Trail', in S. Sumartojo and B. Wellings (eds.), *Nation, Memory, and Great War Commemoration: Mobilizing the Past in Europe, Australia and New Zealand*, Bern: Peter Lang, 291–306.

Sumartojo, S. (2015) 'On atmosphere and darkness at Australia's Anzac Day dawn service', *Visual Communication*, 14(2): 267–288.

Sumartojo, S. and Stevens, Q. (2016) 'Affective atmosphere at the Anzac Day dawn service', in D. Drozdzewski, S. de Nardi and E. Waterton (eds.), *Memory, Place and Identity: Commemoration and Remembrance of War and Conflict*, London: Routledge, 189–204.

Taylor, W. (2005) 'Lest we forget: The Shrine of remembrance, its redevelopment and the heritage of dissent', *Fabrications*, 15(2): 95–111.

Todman, D. (2009) 'The ninetieth anniversary of the Battle of the Somme', in M. Keren and H. Herwig (eds.), *War Memory and Popular Culture: Essays on Modes of Remembrance and Commemoration*, Jefferson, NC: McFarland, 23–40.

Waterton, E. and Dittmer, J. (2014) 'The museum as assemblage: Bringing forth affect at the Australian War Memorial', *Museum Management and Curatorship*, 29(2): 122–139.

Weiler, B. and Yu, X. (2006) 'Understanding experiences of Chinese visitors to Victoria, Australia', CRC for Sustainable Tourism. Available from: http://crctourism.com.au/wms/upload/resources/bookshop/Weiler_ChineseVisitors-VIC.pdf (accessed 16 December 2015).

Wellings, B. and Sumartojo, S. (2017) (eds.) *Commemorating Race and Empire during the First World War Centenary*, Liverpool/Marseille: Liverpool University Press/Presses Universitaires de Provence. *In press*.

Wellington, J. (2016) 'Narrative as history, image as memory: Exhibiting the Great War in Australia 1917–41', in J. Longair and J. McAleer (eds.), *Curating Empire: Museums and the British Imperial Experience*, Manchester: Manchester University Press, 104–121.

Winter, C. (2009) 'The Shrine of remembrance Melbourne: A short study of visitors' experiences', *International Journal of Tourism Research*, 11: 553–565.

Winter, J. (2010) 'Introduction: The performance of the past: Memory, history, identity', in K. Tilmans, F. Van Vree and J. Winter (eds.), *Performing the Past: Memory, History, and Identity in Modern Europe*, Amsterdam: Amsterdam University Press, 11–23.

Witcomb, A. (2013) 'Understanding the role of affect in producing a critical pedagogy for history museums', *Museum Management and Curatorship*, 28(3): 255–271.

Wright, T. (2015) 'Remembrance Day: Hundreds gather at Shrine to pay tribute', *The Age*, 11 November. Available from: www.theage.com.au/victoria/remembrance-day-hundreds-gather-at-shrine-to-pay-tribute-20151111-gkw444.html#ixzz3uBIRQt1K (accessed 13 December 2015).

Ziino, B. (2007) *A Distant Grief: Australians, War Graves and the Great War*, Crawley, WA: University of Western Australia Press.

11 'To leave a wooden poppy cross of our own'

First World War battlefield spaces in the era of post-living memory

Catriona Pennell

On our way to the cemetery, I wondered how I'd feel about seeing the family name I shared with Alfred on his headstone. I asked Peter [tour guide] how people typically react. "It's a mix of emotions; family ties, respect, a new found attachment to it all – everyone handles it differently," he said. "Some break down, some revert into contemplative silence, others want every bit of information I can offer." As we walked into the cemetery, I suddenly felt anxious and slightly sick, as if I were about to meet someone important for the first time. Together we read the inscription: "A.E. ROFF Machine Gun Corps (Inf), 11th July 1917, Age 20". When I saw the white headstone, my reaction surprised me. I felt content, calm, and was glad I'd come to leave a wooden poppy cross of our own.

(Roff 2014: 53)[1]

Although battlefields and other sites of conflict have attracted visitors for over a thousand years, the First World War represents a watershed in the emergence of battlefield tourism (Baldwin and Sharpley 2009: 186). Even as the war raged, travel agencies fielded enquiries from inquisitive members of the public about organised trips to the battlefields of France and Belgium. On 31 March 1915, *The Times* ran a short article entitled 'Trips to Battlefields. No "Conducted Tours" Till The War Is Over' on behalf of three travel companies, including Thomas Cook, announcing that they had not the 'slightest intention of organizing such trips' while the fighting was still in progress (Van Emden 2011: 284).[2]

This all changed once the war was over. Its residual trauma left a deep psychological scar on the imaginations of the belligerent populations involved. It was an unparalleled human tragedy; by 1918, almost all towns and villages throughout Britain had become communities of the bereaved (Winter 1995: 6). In order to try to make sense of the war and its resultant mass death, the post-1918 generation created social memories which served to honour and remember the dead. Thousands of memorials were created across the world in villages, cities as well as on the former battlefields themselves (Winter 2009: 607).[3] Many initiatives were driven by official public processes of honouring the dead; in Britain the main examples are the Cenotaph in Whitehall and the Tomb of the Unknown Soldier at Westminster Abbey (both unveiled on 11 November 1920) (Winter 1995).[4]

Many more were the result of an unofficial, often private, desire to remember individual family members or groups of men connected by a community of place, work, faith and/or sport. As the majority of the dead were not repatriated home, the battlefields became sites of pilgrimage and tourism combining an interesting mix of public and private desire to remember. The desire to visit places associated with the war was shared by people in many ex-combatant nations including Australia, Canada, France, Germany and America. The British travel agency, Thomas Cook, organised the first trips in 1919; in that year alone, 60,000 people visited the battlefields (Seaton 2000: 63). The motives of these initial visitors ranged from morbid curiosity to a desire to commemorate lost loved ones. According to Lloyd, 'the appeal of these sites to tourists and pilgrims is indicative of the pervasive presence of the war and the sense of loss which it engendered in the fabric of life in many of these countries' (Lloyd 1998: 2). In the interwar period a range of battlefield guides were published, most famously by Michelin (Dunkley et al. 2011: 862). Although after the Second World War the Western Front became a largely forgotten landscape, it once again become a popular destination for British visitors – despite the war receding further into the past, with no known surviving veterans alive today (Iles 2008: 138).[5] While harder to gauge visitor numbers at memorials and battlefields, those that have accompanying visitor centres give some indication of the level of present-day foot traffic. In 2009/10, 133,987 people visited the Thiepval Memorial and Visitor Centre (France); in the following year, 198,542 people visited the In Flanders Fields Museum in Ypres (Belgium) (Miles 2013: 223).

As one recent commentator concludes: 'Reading books and watching programmes provides background, but nothing sheds more light . . . than visiting the sites' (Andres 2015). People can now travel to the battlefields independently using one of the many published guidebooks (e.g. Holt and Holt 1997, 2008) or digital apps.[6] Alternatively, people can join an organised coach tour, which, according to Miles (2013) dominates the British Western Front tourism market and places the tour guide in a pivotal interpretive role.[7] As the centenary unfolds, visiting the Western Front has taken on increased significance with a number of agencies organising special tours to coincide with particular anniversaries (Andres 2015). British military service charities are utilising the centenary (particularly the Battle of the Somme) as a fundraising opportunity with sponsored walks, marathons and bike rides taking place across the Western Front.[8] Representatives from key stakeholders such as the Commonwealth War Graves Commission (CWGC) and the International Guild of Battlefield Guides testify to the centenary stimulating unprecedented visitor numbers to the battlefields.[9] Visitor numbers to West Flanders (Belgian Westhoek) have risen by over 140 per cent between 2006 and 2014.[10] The battlefields themselves have taken centre stage in a number of official state-sponsored commemorative activities. In 2016 alone, the UK government hosted a commemorative service at Thiepval Memorial on the Somme (1 July), the French authorities at Verdun (29 May) and the Australian government at Fromelles and Pozières (19 July). Further afield, Australian battlefield tourism companies were already fully booked for the 2015 Dawn Service at Gallipoli as early as 2012 (McKay 2013: 1). On 25 April 2015, it was reported that more than 10,000 Australians and New Zealanders stood together on the shores of the peninsula at a formal ceremony to honour the Anzac troops who fought and fell in Turkey a century ago.[11]

Over the 100 years since the end of the war, numerous changes have occurred within British society. With this in mind, this chapter seeks to examine and compare the experience of First World War battlefield tours by British visitors from the 1920s to the present day in order to ascertain how these visits have evolved, particularly in regards to the difference between pilgrimage and educational tour. It first engages with definitions of both 'pilgrim' and 'tourist' in the context of visitor motivations when travelling to the battlefields of the Western Front before demonstrating that the former is the most appropriate way of describing visitors in the immediate post-war and interwar period. As the temporal distance from the event itself increased, the type of visitor to France and Flanders tended to fall more squarely in the category of educational tourist (which contained implicit elements of a commercialised, pleasure-seeking activity) although the sense of honouring the dead always remained explicit. Prominent case studies from the disciplines of tourism research and historical geography in the 1990s and 2000s are then reviewed as a platform from which to offer initial insights into battlefield tourism in the era of the centenary, with a particular focus on the British government-funded First World War Centenary Battlefields Tour Programme (FWWCBTP).[12]

Pilgrim or tourist?

In order to better understand the reasons why people have visited First World War battlefield sites, it is necessary to try and distinguish between the two most common forms of visitation: pilgrimage and battlefield tours. Pilgrimage is best understood as a journey or search for moral or spiritual significance.[13] In the case of battlefield pilgrimage, it can be interpreted as an act of mourning and/or remembrance, the focus being on the spiritual value of visiting a grave. While tourism more broadly revolves around ideas of entertainment and leisure, the purpose of a battlefield tour, however, is primarily (although not solely) about education.[14] It is an attempt to understand what happened at a particular site and why. As explored in this chapter, in practice the distinction between battlefield pilgrimage and tour is blurred. Visitors come to the Western Front to follow their ancestors' story whilst others set about tracing the experience of those unrelated to them. This phenomenon of 'vicarious pilgrimage' is a common feature of Western Front tourism that underscores the intensely personal nature of visits to these places (Dunkley et al. 2011; Miles 2013). While someone might visit a site as an act of pilgrimage with no interest in the background history, just as someone might visit the same place without any element of pilgrimage or homage to the dead, it is also plausible that someone who visits the battlefields can alternate between the position of pilgrim and tourist over the course of the trip (Scates 2006; Baldwin and Sharpley 2009: 190–191). Visits are therefore best understood as a 'hybrid experience' where motivations for making the journey are multifaceted and fluctuate (Winter 2011: 173).

Interwar pilgrimage

In the conflict's immediate aftermath, it was clear that to undertake a pilgrimage was the primary reason why grieving relatives were making the journey across the Channel. Owing to the scale of the losses and the framing of the war as a Manichean struggle between good and evil where soldiers were understood to be sacrificing their lives for the survival of their nation, the First World War battlefields soon became understood in the immediate post-war period as sacred places to be visited by pilgrims rather than tourists (Lloyd 1998: 26). 'Tourist', in the context of the Western Front at this time, was a disparaging term; referring to curious and disrespectful day-jaunters, 'sallying out from their comfortable hotels in fast motor-cars to "do" the battlefields and pick up over-priced fake souvenirs' (Pegum 2008: 218). Pilgrims, on the other hand, were bereaved relatives of dead or missing soldiers travelling sombrely and reverently to pay their respects in whatever way they could. Indeed, any suggestion of a visit being for a reason other than mourning was frowned upon by politicians and public commentators at the time. Although unsuccessful, an attempt was made in the House of Commons to exclude tourists from the former front lines – until the relatives of the dead had visited their graves – through fear of bad behaviour (Lloyd 1998: 40).[15] Efforts were made in the published material that accompanied tours to indicate the appropriate moral behaviour of visitors; according to one 1919 Michelin guide visitors were encouraged to see their trip as a pilgrimage, not 'merely a journey across a ravaged land' (Iles 2008: 141). Pegum (2008) highlights a third category of visitor in the post-war period: the ex-servicemen who were neither tourist nor pilgrim but a complex mixture of both and who often found, on their return to the Western Front, an unfamiliar landscape.

Battlefield pilgrimages at this time tended to be dominated by two groups: ex-servicemen and bereaved relatives and, within these groups, limited to the middle and upper classes who could afford to pay at least £4 per person to join a tour or pilgrimage (Lloyd 1998: 48). Welfare organisations such as the Ypres League and the British Legion assisted people with journeys in the form of organised tours (Vanneste and Foote 2013). The Western Front battlefields attracted the vast majority of visitors from Britain, but a surprising number made the journey to Gallipoli (perhaps a more romantic evocation of the war compared to the misery and losses of the Western Front) and the war cemetery in Jerusalem (reaffirming a sense of religious crusade). In the immediate aftermath of the war the levels of visitors to these sites reflected a more general obsession with the war, which had pervaded every aspect of peoples' lives. This would not last and the mid-1920s saw a decline in battlefield travel. However, the years of the 'war books boom' in the late 1920s and early 1930s saw resurgence. While 'pilgrimage' dominated the immediate aftermath, 'tourism', more clearly, took over in the 1930s. Though the onset of global depression tempered ability to travel, visitor numbers remained relatively high. Even with the deteriorating international situation, people still returned to these sites, some to prepare for future conflict, others as part of an effort to strive for peace (Lloyd 1998: 130–131) adding an interesting third trope of 'peace tourism' to battlefield visitor motivation in this period.[16]

By the 1930s, the practice of battlefield pilgrimage had evolved from being simply the personal journey of a bereaved relative to a public act of commemoration. The war had enabled the intertwining of family and national histories (Winter 2006: 2) and the nation stood 'with the pilgrim at the graveside' (Baldwin and Sharpley 2009: 194). Pilgrimages were at once an individual and public experience, which united groups and the nation in remembrance. An important example of this, as discussed by Lloyd (1998), is the Royal British Legion (RBL) pilgrimage to the Western Front of 1928. Between the 3 and 8 August 1928, around 11,000 people took part, vastly overshadowing the numbers that attended the opening of the Menin Gate memorial almost one year earlier.[17] There was a widespread feeling that pilgrims were engaged in something other than an ordinary journey. Non-participants waved the pilgrims off from departure points in the United Kingdom. Delegations were sent to lay wreaths at the tombs of the Unknown Soldier in Paris and Brussels as well as sites associated with their group's background or where they were staying.[18] The pilgrimage culminated in a large memorial service held at the Menin Gate (see Figure 11.1 below) where pilgrims were addressed by the Archbishop of York before marching past the Prince of Wales. The BBC also broadcast the service across all its radio stations.

Figure 11.1 Members of the Royal British Legion pilgrimage to the battlefields, attending the service at the Menin Gate, August 1928.

Source: © Royal British Legion Buxton Branch, courtesy of Mr R. N. Nicol.

The character of the pilgrimage – as desired by the organisers – was pious, digni-
fied and helped to underscore how the pilgrimage was about paying homage to
the dead rather than any sense of celebrating Britain's victory or enjoying a nice
holiday. As participation was expensive, there was an additional awareness that
those people making the trip had a 'duty' in place of those who could not afford to
travel. The language of the 1928 guide is indicative of this mood:

> We Pilgrims are going to offer homage to the dead of the British Empire . . .
> the great majority of whom lie buried in those beautiful cemeteries in France
> and Flanders. And in this act itself we shall be bringing consolation to the
> widows and orphans, the mother who has lost her boy, the maiden the man
> of her choice. Those who are not able to come themselves will learn from
> those who have seen, how comforting is the serene beauty of those white
> headstones.
>
> (Harter and Gavin 1929 cited in Baldwin and Sharpley 2009: 192)

1945 to the 1990s

The Second World War, with the closing of mainland France to visitors, prevented
visits to the old Western Front. Its end saw a generation focus their attention on
this more recent war (Seaton 2000: 63). Visits to the First World War battlefields
rapidly declined during the 1950s and 1960s, as the Western Front became 'a
largely forgotten landscape' (Saunders 2001: 45). The author and tour guide,
Martin Middlebrook, recalled that during his first trip in 1967, he and his friend
'met no other visitors on our travels and there seemed to be no organised tours'
(Middlebrook and Middlebrook 1991: 2 cited in Iles 2008: 142).

From the 1970s onwards, there was a resurgent interest in the battlefields. This
was influenced by a number of factors: the fiftieth, sixtieth and seventieth anniversaries of the start of the First World War (in 1964, 1974 and 1984 respectively); the
broadcasting of the BBC's 26-episode documentary series *The Great War* (1964);
the publication of an increasing number of books, including Middlebrook's influential *First Day of the Somme* (1971); the formation of the Western Front Association, inaugurated on 11 November 1980; and the promotion by several commercial
organisations of battlefield tours. The early 1970s saw some 50,000 visitors to the
Western Front each year. By 1974 this had increased to 250,000 (Saunders 2001:
45). Similarly, personal enquiries concerning the whereabouts of a grave made to
the CWGC rose from 1,500 a year in the mid-1960s to 8,000 in 1980 and 28,000
in 1990 (Walter 1993: 63). By the war's eightieth anniversaries, in the 1990s,
interest had increased significantly, particularly among British young people
through schools and educational trips. A whole industry of guided tours and specialist publishing, along with early forms of Internet resources, appeared to support this growing tourist activity (Seaton 2000: 64). The rise of digital resources
has been crucial to democratising access to archival material that, in turn, might
stimulate the desire to visit the battlefields (Fabiansson 2004). Furthermore, this
increased interest in tracing relatives or local figures must be placed within the

broader phenomenon of family history that has captivated global audiences and influenced flagship television programmes such as the series *Who Do You Think You Are?* (Wallis 2015), as well as Internet subscription sites like Ancestory™, the world's largest genealogy website.[19] The recent popularisation of battlefield archaeology – moving past memorialisation of the war towards an attempt to present the experience of it – stands as another contributory factor (Robertshaw and Kenyon 2008; Price 2005).

Studies of visitor perspectives during this period coincided with an emerging academic research interest in 'dark tourism' (Foley and Lennon 1996) and 'thanatourism' (Seaton 1996); a form of travel to landscapes associated with death and violence, including battlefields. In April 1998, Seaton conducted an exploratory study of a conducted coach party touring the First World War battlefields of the Somme and Flanders (Seaton 2000). In summary, the motivations for travelling to the Western Front and the responses to the battlefield tour experience were dependent on the social profile of each visitor. Out of 29 participants, Seaton identified three separate groups: a student party (travelling as part of their academic studies); recreational tourists (people who had joined the party as a weekend leisure trip); and military enthusiasts (travelling as part of a local military society). All showed differences in commitment and stamina in pursuing the tour programme, the latter group being the most dedicated to all aspects of the tour content that related to the war (i.e. choosing to visit a military museum in Albert rather than having a coffee break in a local café). In terms of tour content, as with any organised coach party, the programme was constructed. Out of 36 stops made during the three-day tour, 31 were at war memorials and military cemeteries, occupying more than 75 per cent of the tour. Thus, the construction of the visit and meanings attached to it were 'overwhelmingly structured around these commemorative memorial sites' constructed and managed by the CWGC (Seaton 2000: 67). Interest as displayed by these visitors was person-centred and compassionate as visits to the cemeteries and memorials 'produced reactions of homage to the dead' rather than any kind of 'voyeuristic necrophilia' (Seaton 2000: 75).

Battlefield visitors at the ninetieth anniversary

According to a 2006 survey commissioned by the RBL (that sampled 1,000 respondents across the United Kingdom) 28 per cent of people had visited a battlefield or war memorial overseas. A series of significant anniversaries, combined with ease and affordability of travel and a surge in the popularity of history, particularly family history, led to growing demand for battlefield tours. The popularity of battlefield visits at the war's ninetieth anniversary was able to support at least ten British travel firms (Iles 2008). Annual visitors to the Westhoek in Flanders were estimated, in 2008, to number 326,900 (Vandaele and Monballyu 2008).

Up until this point, the majority of studies about battlefield tourists were qualitative – like Seaton's work earlier – providing rich and detailed analyses of visitor motivations and experiences from a small group sample, rather than information on broad-scale visitation trends to a particular location. To rectify this lacuna, in July 2009, Caroline Winter sampled visitors to the Belgian town of Ieper (known as Ypres particularly amongst British visitors). The findings were based on 145 completed questionnaires (distributed at the Visitor Information Centre located in the Cloth Hall, which houses the *In Flanders Fields* museum) primarily from British (31.5 per cent), Belgian (25.9 per cent) and French (14 per cent) visitors, as well as small numbers of Germans, Australians, Americans, Canadians and one person each from Poland, Denmark, South Africa and Czechoslovakia [sic]. Most (78.6 per cent) were accompanying friends and family, and only small proportions of the sample were on a tour (11 per cent) or travelling alone (9 per cent). The majority of British visitors sampled travelled because of a desire to learn more about the battlefields (53.3 per cent) or because of a direct connection (e.g. through a family member's grave) with a particular battlefield (36.6 per cent) (Winter 2011: 168–171).

Winter's study destabilises the distinction between pilgrim and tourist. Most of the sample would be appropriately described as tourists: visiting battlefields and graves was not their sole reason for travel; they sought leisure experiences on top of their visits to such sites; and travelled as part of a longer holiday trip. Yet at the same time, remembrance activities – visiting war memorials, cemeteries and battlefields, as well as learning more about the war – had high importance for many. There existed a strong requirement to fulfil some kind of cathartic or emotive experience; it had to carry 'meaning' for the participant(s). While only a small proportion of visitors could be seen as pilgrims, this left a significant group who appeared to be tourists, but who shared most of the characteristics of pilgrims. Equally, many self-identified pilgrims exhibited similar characteristics of tourists (Winter 2011: 173). Thus although Baldwin and Sharpley (2009) suggest that a line can be drawn between those 'tourists' who travelled to battlefields to understand and learn about the war and 'pilgrims' who travelled to pay respect to the dead, this dichotomy may obscure and deny the experiences of some visitors to the Western Front (Winter 2011: 165). As Seaton (1996), Gatewood and Cameron (2004), and Iles (2008) have all illustrated, those without a direct link to the dead can still have a deep and meaningful experience.

Pilgrims in school uniforms: battlefield tours and the UK government response to the centenary

There is a growing anxiety that as the temporal distance between the current generation and the war extends, emotional and invested interest amongst the general public will wane. While there is plenty of evidence to indicate a sustained and perhaps growing interest in the war at the time of the centenary anniversaries amongst the post-1945 'baby boom' generation, there is less information regarding the responses of younger people (Winter 2011). A concern about carrying the memory of the war forward may help to explain why much of the state-sponsored centenary activity in the United Kingdom (and other combatant nations, such as Australia and New Zealand) has been focussed on young people and education (Pennell and Sheehan 2016).

The cornerstone of the UK government's commemorative activity is the FWW-CBTP. This £5.3million programme is funded by the Department for Education (DfE) and the Department for Communities and Local Government (DCLG), as part of the national centenary commemorations. It is designed to provide the opportunity for a minimum of two students and one teacher from every state funded secondary school in England to visit the Western Front battlefields between 2014 and 2019.[20] The tours are led by members of the International Guild of Battlefield Guides and each coach is accompanied by a serving member of the British armed forces. Although the tour content varies slightly, all students visit the major memorial sites including the Menin Gate, Tyne Cot cemetery, Thiepval Memorial to the missing on the Somme, and the German war cemetery at Lange-mark, as well as at least one museum in Belgium.[21]

Visitors to the Great War battlefields are not simply passive observers. As part of a growing scholarship on the embodied and performative nature of tourist practices (Edensor 1998, 2000; Kirshenblatt-Gimblett 1998; Desmond 1999; Coleman and Crang 2002; Crouch and Lübbren 2003; Franklin 2003) it is important to acknowledge the degree to which tourists, travelling as part of an organised group, are directed and supervised on their visits and the impact this has their experience (Iles 2008: 140). In terms of the constructed nature of the FWWCBTP, the following elements need be considered: the objectives of the government-sponsored programme; the pre-departure preparatory information circulated; the choice of sites visited; and the role and rhetoric of the guide(s) who led and managed the tour.

The official objectives of the FWWCBTP are to deepen participants' under-standing of the First World War through personal connections and to create an enduring legacy for future generations of pupils.[22] Evidently, the lines between education and commemoration are blurred. Throughout the tour programme, start-ing with the pre-departure training for teachers who accompany their pupils, and including the workbook and pre-departure briefings, emphasis is placed on the visit being as much about learning more about the war as paying respects to the dead. Historical understanding is blended with emotive remembrance. The major-ity of stops are at CWGC-managed cemeteries and memorials, thus underscoring the symbolic, commemorative nature of the visit. The pupils engage in research activities that are a mix of education and homage to the dead, such as investigat-ing the story of a dead soldier 'adopted' by the pupils at the start of the tour that is connected to their local area; when they find 'their soldiers' grave, they are encouraged to lay a wooden cross with their names on and school name provided by the RBL (see Figure 11.2 below). There are also strongly ritualistic aspects of the programme. Participants attend the Last Post Ceremony at Menin Gate (two pupils from each coach are chosen to lay a wreath accompanied by a serving Brit-ish soldier) as well as taking part in a closing wreath-laying ceremony at Tyne Cot where, in tours observed in 2015 and 2016, the fourth stanza of Laurence Binyon's poem 'For The Fallen' was read aloud with the final line 'we will remember them' repeated in unison by the entire tour party, heads bowed and followed by a min-ute's silent contemplation.[23]

Figure 11.2 Participating FWWCBTP student places a wooden cross at a panel on the
    Menin Gate, March 2015.

Source: © First World War Centenary Battlefield Tours Programme, courtesy of Simon Bendry.

Naturally, what the tour participants are presented with and what they take
from the experience are not necessarily the same. The tour organisers may place
emphasis on education and commemoration, but each participant has their own
individual ideas and responses. Tour observations, surveys, interviews and focus
groups all provide an opportunity to unravel what is communicated to the participants and what they receive. Students who participated in the Menin Gate
ceremony exhibited behaviour similar to pilgrims and talked about its emotional
weight. These sacred places are experienced as well as seen; there is an embodied
semi-spiritual interaction with the surfaces (the battlefields and memorials) that
the students visit during the course of their tour. That said, as soon as the ceremony
was over, they all hurried to the town's chocolate shops to stock up on souvenirs
reaffirming the fluctuating identity of 'pilgrim' and 'tourist' discussed earlier.

It is important to note, however, that while the FWWCBTP participants are
volunteers, in that they have put themselves forward to be involved, they are
selected to go on the tour by their teachers. This can be for a variety of reasons
including socio-economic background, no previous opportunity to go abroad with
school, academic performance, personal interest in the First World War, or as a
reward for good behaviour. Numerous teachers testified they had chosen students
whom they knew would behave well and be easy to manage on a trip where their
students were acting as 'external ambassadors' and representing the school in a

government-funded programme. The manner in which the FWWCBTP is constructed affords limited room for a pupil to act as a pleasure-seeking tourist or to ask critical questions about what exactly is being remembered, and why. Might questions be conceived as disruptive and disrespectful, insulting the memory of past and present service personnel? Asked how she might respond to a member of her coach party who felt remembering the First World War glorified conflict, one pupil who participated in a spring 2015 FWWCBTP tour explained:

> I just really disagree with that viewpoint . . . it's like walking into a church and you know saying that you love the devil and you hate god and everything. It's *not* appropriate. It's OK to be an atheist completely I would not disagree with that at all but not in that context and I think it may have been more acceptable outside that situation like where we were staying or on the coach but even then we were talking about there's a lot of peer pressure because the tour was *to remember* and to learn about that you know not many people there are going to put their hands up and agree with you because that's not the purpose of going.[24]

For this young person, expressing pacifist views in the context of the battlefield tour was tantamount to being a non-believer, a heretic in the church of national sacrifice. While there is still work to be done in investigating what cultural memory messages young people are taking from these tours, there appears to be an onus on remembrance; respect and honouring the dead are requirements for participation. This is not dissimilar to expectations in the 1920s and 1930s that Western Front pilgrims would behave in a morally acceptable way. While to some extent present-day battlefield tourism is making a hesitant shift towards a new purpose of utilising the landscape as a means towards reconciliation, in other ways it is deepening traditional channels of British war remembrance: unquestioning reverence and homage to the dead soldier.

Conclusion

Visiting the battlefields of the Western Front has been a popular pastime for Britons since the end of the First World War. However, it is more difficult to ascertain individual motivation, in particular to distinguish whether visitors should be categorised as pilgrims (who wish to pay homage to the dead) or tourists (albeit travelling to these sites for the purpose of education rather than entertainment).

It might be assumed that today's young people – owing to the temporal distance between their lives and the conflict – might be less interested in visiting the battlefields and that if they do, it would be for vastly different reasons than visitors in the 1920s and 1930s.[25] Granted, there is increased commercialisation of these heritage sites that cannot be overlooked, with further concerns raised about what will happen after the centenary ends. Will it result in a 'Disneyfication' of the landscape in order to keep attracting visitors whose connection to that particular past is tenuous at best? The challenge for stakeholders such as the CWGC is to build upon the interest generated by the centenary and thus ensure it has a legacy beyond the commemorative events being held to mark it.[26]

What can we ascertain about visitor motivation and responses now, as the centenary unfolds? The young people involved in the FWWCBTP are an excellent case study. In due course, I (along with colleagues in the United Kingdom, Australia, Canada and New Zealand) will be contributing detailed analysis and material to precisely the question of why young people visit the battlefields of the Western Front and what they take from that experience, as well as broadening these questions out beyond the battlefields to youth interaction with the centenary via formal and informal education. Certainly, a sense of heightened anxiety about what will happen to the memory of the First World War at its one hundredth anniversary has fuelled official state-sponsored commemorative activity focussed on youth participation. The FWWCBTP is the cornerstone of this in the United Kingdom. In part, as a government-funded initiative, perhaps it tells us more about political incentives to remember, rather than any sense of what young people are actually doing participating in the tour.[27] Initial data gathered from participants in the 2015 tour groups indicate that participation is driven primarily by a desire to remember and pay homage to the dead, rather than as an opportunity to travel abroad or get time off school. But is this spiritual element being instructed from above (teachers, tour guides, families) rather than felt spontaneously from within? While some students are looking for a relative, most are researching a stranger (where the connection is either through their locality or their school). It is possible that students are responding to the location of this soldier's grave with respect and homage because that is the only appropriate way to respond according to societal norms and the pressure of representing their school.[28]

The integration of a visit to Langemark German war cemetery and the active involvement of many FWWCBTP participants in the Gone West 'Coming World Remember Me' sculptural art project are attempts at integrating messages of peace and reconciliation.[29] Building on developments in the late 1930s as well as more prominent examples in the 1990s, does this imply a third category of visitor to the Western Front beyond 'pilgrim' and 'tourist'? I think we have to be careful not to exaggerate this evolution.[30] Homage to the dead, respect, and pride in a British national identity remain prominent ways in which FWWCBTP participants make sense of their experience. In many ways, there are more commonalities than differences between this government-funded programme and that of the RBL tour in 1928. Rather than encouraging participants to 'surf the barriers of national memory cultures and gain a broader sense of the conflict' (Reynolds 2013: 433) there is a risk that traditional enclaves of 'appropriate' and state-managed remembrance are being deepened by creating compliant pilgrims in school uniforms.

Notes

1   Ali Roff, a journalist with *Psychologies Magazine*, documented her family visit to her great uncle's grave in Armentières, France in 2014. She was following in the footsteps of Alfred's parents, James and Emily Roff, who made a pilgrimage to their son's grave with his brother and sister-in-law immediately after the war.

2 'Trips to Battlefields. No "Conducted Tours" Till the War Is Over', *The Times*, 31 March 1915, p. 5.

3 There is a vast literature on the topic of First World War remembrance evidenced in academic bibliographic resources; Kovacs and Osborne (2014) and the International Society for First World War Studies collaborative online bibliography (see www. firstworldwarstudies.org/bibliography.php accessed 10 June 2016). Seminal examples of historical geography scholarship include: Heffernan (1995), Foster (2004), Gough (2004), Johnson (2008) and Switzer and Graham (2010).

4 After the war, commemoration patterns varied from country to country. The victorious Allies soon developed a series of similar 'official' traditions in commemorating the conflict and honouring the dead, such as national days of remembrance, the dedication of tombs to unknown soldiers, and the erection of monuments.

5 Harry Patch, the last British veteran who served in the trenches, died on 25 July 2009 aged 111. Florence Green, a British citizen who served in the Women's Royal Air Force, died on 4 February 2012, aged 110.

6 For instance, see www.greatwar.co.uk/trip-info/battlefield-app.htm (accessed 18 March 2016).

7 The distinction between independent travel and organized tour is purely logistical; I acknowledge that organized tours do not deny the space for personal inward and intimate experiences for individual participants.

8 See The Western Front *Via Sacra* Walk – 2016: www.viasacrawalk2016.org.uk/ and Help for Heroes Big Battlefield Bike Ride 2016: The Western Front: www. helpforheroes.org.uk/get-involved/challenges/cycling-for-charity/big-battlefield-bike-ride-2016/ (accessed 23 June 2016).

9 Email correspondence between the author and Mike Peters, Chairman of the International Guild of Battlefield Guides (27 April 2016); email correspondence between author and Peter Francis, Media and Marketing Manager, CWGC (3 May 2016). However, both acknowledged the negative impact the terrorist incidents in Paris (November 2015) and Belgium (March 2016) have had on visitor numbers.

10 Persconferentie Herdenkingstoerisme Westhoek Bezoekerscijfers 2014 (12 January 2015). Kindly provided by Peter Francis.

11 www.theguardian.com/news/2015/apr/25/anzac-day-in-gallipoli-dawn-service-takes-place-on-100th-anniversary (accessed 18 March 2016).

12 The tours are organized by the Institute of Education (IoE) in partnership with Equity Travel. They are funded by the Department for Education (DfE) and the Department for Communities and Local Government (DCLG). See www.centenarybattlefieldtours. org (accessed 19 May 2016).

13 www.britannica.com/topic/pilgrimage-religion (accessed 19 May 2016).

14 That is not to overlook the sociability of trips, including meeting new, possibly like-minded people and enjoying the cuisine and fine wine of France and Belgium.

15 J.H. Franklin, an American visiting the battlefields in 1919, found them to be perilously 'bestrewn with unexploded shells and other missiles' and 'not yet a land for tourists' (Franklin 1919: 64 cited in Pegum 2008: 219).

16 Seen as a place to promote peace, this instrumentalisation of the battlefields became popular in the 1990s context of peacekeeping efforts such as the Good Friday Agreement (e.g. the Island of Ireland Peace Park at Messines, Belgium). Jansen-Verbeke and George (2013) have discussed reconfiguring the battlefields as sites of 'peace tourism' during the centenary period.

17 Although impossible to get exact numbers, the fact that all hotel accommodation in Ypres was booked on 7 August suggests that many other people joined the pilgrims. The *Daily Express* estimated 50,000 people were present in addition to the pilgrims (Lloyd 1998: 155).

18 For example, both the Irish Free State party and the Ulster party laid wreaths at the memorial to the Ulster Division at Thiepval (Lloyd 1998: 154).

19 See www.ancestry.co.uk (accessed 9 June 2016).
20 It is intended that, by 2019, 12,000 young people will travel to the Western Front on this programme – more than the 1928 RBL tour. Granted, the FWWCBTP is achieving this statistic over the course of five years, as opposed to a single trip, but it is worth asking whether the DfE and DCLG are conscious of the new record they intend to set for battlefield pilgrimage?
21 At the time of writing, this is either the *In Flanders Fields* museum and/or the Passchendaele Memorial Museum in Zonnebeke.
22 www.centenarybattlefieldtours.org/about-us/further-information-and-acknowledgements/ (accessed 7 April 2016).
23 The relationship between performative ritual and the formation of national and/or religious identity can be found in examples, such as the Methodist Sunday school parades in west Cornwall in the mid-nineteenth to twentieth centuries (Harvey et al. 2007).
24 FWWCBTP Spring 2015 participant comments in July 2015 focus group.
25 There is the often-peddled accusation that 'the youth of today' have no respect for their forefathers' wartime sacrifice, although I (and others) would posit this more of a media construction than a reflection of reality. See, for example, Southern, (31 May 2016).
26 Email correspondence between author and Peter Francis (3 May 2016); email correspondence between author and Major Tonie Holt (22 April 2016).
27 As Foster states 'commemoration for the purposes of present politics is nothing new' (2001: 219). Harvey (2017) and Mycock (2014) both explore the way the British government is instrumentalising First World War memory to promote certain present-day agendas and policies particularly regarding the British Army and the campaign in Afghanistan.
28 For more on the subject of 'deathscapes' being simultaneously burial grounds and morally instructional spaces see Deering (2010).
29 This aims to create 600,000 clay sculptures by 2018 – each representing a soldier and spanning 50 nationalities that died on Belgian soil during the war. The sculptures will be unveiled in the no man's land of the frontline around Ypres in 2018. See www.comingworldrememberme.be/en/the-land-art-installation (accessed 10 June 2016).
30 Caution seems necessary particularly in the British context. Centenary-inspired collaborative educational projects involving French and German schoolchildren took place at Verdun in 2016 that sought to transcend national memorial cultures, as a place of European memory. See 'Le concours "Verdun – lieu de mémoire européen ?", http://centenaire.org/fr/espace-pedagogique/mobilisation-de-la-communaute-educative/le-concours-verdun-lieu-de-memoire (accessed 23 June 2016).

Bibliography

Andres, T. (2015) 'The best First World War battlefield tours for 2016', *Telegraph*, 4 November. Available from: www.telegraph.co.uk/travel/tours/The-best-First-World-War-battlefield-tours-for-2016/ (accessed 18 March 2016).

Baldwin, F. and Sharpley, R. (2009) 'Battlefield tourism: Bringing organised violence back to life', in R. Sharpley and P.R. Stone (eds.), *The Darker Side of Travel: The Theory and Practice of Dark Tourism*, Bristol: Channel View Publications, 186–206.

Coleman, S. and Crang, M. (eds.) (2002) *Tourism: Between Place and Performance*, New York: Berghahn Books.

Crouch, D. and Lübbren, N. (eds.) (2003) *Visual Culture and Tourism*, Oxford: Berg.

Deering, B. (2010) 'From anti-social behaviour to X-rated: Exploring social diversity and conflict in the cemetery', in A. Maddrell and J. Sidaway (eds.), *Deathscapes: Spaces for Death, Dying, Mourning and Remembrance*, Farnham: Ashgate, 75–93.

Desmond, J. (1999) *Staging Tourism: Bodies on Display from Waikiki to Sea World*, Chicago, IL: University of Chicago Press.

Dunkley, R., Morgan, N. and Westwood, S. (2011) 'Visiting the trenches: Exploring meanings and motivations in battlefield tourism', *Tourism Management*, 32: 860–868.

Edensor, T. (1998) *Tourists at the Taj: Performance and Meaning at a Symbolic Site*, London: Routledge.

Edensor, T. (2000) 'Staging tourism: Tourists as performers', *Annals of Tourism Research*, 27: 59–81.

Fabiansson, N. (2004) 'The internet and the Great War: The impact on the making and meaning of Great War history', in N.J. Saunders (ed.), *Matters of Conflict: Material Culture, Memory and the First World War*, Routledge: London, 166–178.

Foley, M. and Lennon, J.J. (1996) 'JFK and dark tourism: A fascination with assassination', *International Journal of Heritage Studies*, 2(4): 198–211.

Foster, J. (2004) 'Creating a temenos, positing "South Africanism": Material memory, landscape practice and the circulation of identity at Delville Wood', *Cultural Geographies*, 11: 259–290.

Foster, R.F. (2001) *The Irish Story: Telling Tales and Making It Up In Ireland*, London: Penguin.

Franklin, A. (2003) *Tourism: An Introduction*, London: Sage.

Franklin, J.H. (1919) *In the Track of the Storm: A Report of a Visit to France and Belgium, with Observations Regarding the Needs and Possibilities of Religious Reconstruction in the Regions Devastated by the World War*, Philadelphia, PA: American Baptist Publications Society.

Gatewood, J. and Cameron, C. (2004) 'Battlefield pilgrims at Gettysburg National Military Park', *Ethnology*, 43: 193–216.

Gough, P. (2004) 'Sites in the imagination: The Beaumont Hamel Newfoundland Memorial on the Somme', *Cultural Geographies*, 11: 235–258.

Harter, J. and Gavin, L. (1929) *The Story of an Epic Pilgrimage*, London: British Legion.

Harvey, D.C. (2017) 'Critical heritage debates and the commemoration of the First World War: Productive nostalgia and discourses of respectful reverence during the Centenary', in H. Silverman, E. Waterton and S. Watson (eds.), *Heritage in Action: Making the Past in the Present*, New York: Springer Press, 107–120.

Harvey, D.C., Brace, C. and Bailey, A.R. (2007) 'Parading the Cornish subject: Methodist sunday schools in west Cornwall, c. 1830–1930', *Journal of Historical Geography*, 33(1): 24–44.

Heffernan, M. (1995) 'For ever England: The Western Front and the politics of remembrance in Britain', *Ecumene*, 2: 293–323.

Holt, T. and Holt, V. (1997) *Major and Mrs Holt's Battlefield Guide to the Ypres Salient*, Barnsley: Pen and Sword.

Holt, T. and Holt, V. (2008) *Major and Mrs Holt's Battlefield Guide to the Somme*, Barnsley: Pen and Sword.

Iles, J. (2008) 'Encounters in the fields: Tourism to the battlefields of the Western Front', *Journal of Tourism and Cultural Change*, 6(2): 138–154.

Jansen-Verbeke, M. and George, W. (2013) 'Reflections on the Great War centenary: From warscapes to memoryscapes in 100 years', in R. Butler and W. Suntikul (eds.), *Tourism and War*, London: Routledge, 273–287.

Johnson, N. (2008) *Ireland, the Great War and the Geography of Remembrance*, Cambridge: Cambridge University Press.

Kirshenblatt-Gimblett, B. (1998) *Destination Culture: Tourism, Museums, and Heritage*, Berkeley, CA: University of California Press.

Kovacs, J.F. and Osborne, B.S. (2014) *A Bibliography: The Great War (1914–1918)*, Halifax, NS: World Heritage Tourism Research Network, Mount Saint Vincent University. Available from: www.whtrn.ca (accessed 17 June 2016).

Lloyd, D.W. (1998) *Battlefield Tourism: Pilgrimage and the Commemoration of the Great War in Britain, Australia and Canada, 1919–1939*, Berg: Oxford.

McKay, J. (2013) 'A critique of the militarisation of Australian history and culture thesis: The case of Anzac battlefield tourism', *Journal of Multidisciplinary International Research*, 10(1): 1–25.

Middlebrook, M. (1971) *The First Day on the Somme: 1 July 1916*, London: Allen Lane/The Penguin Press.

Middlebrook, M. and Middlebrook, M. (1991) *The Somme Battlefield: A Comprehensive Guide from Crecy to the Two World Wars*, London: Penguin.

Miles, S. (2013) 'From Hastings to the Ypres salient: Battlefield tourism and the interpretation of fields of conflict', in R. Butler and W. Suntikul (eds.), *Tourism and War*, London: Routledge, 221–231.

Mycock, A. (2014) 'The politics of the Great War centenary in the United Kingdom', in S. Sumartojo and B. Wellings (eds.), *Nation, Memory and Great War Commemoration: Mobilizing the Past in Europe, Australia and New Zealand*, Oxford: Peter Lang, 99–118.

Pegum, J. (2008) 'The old front line: Returning to the battlefields in the writings of ex-servicemen', in J. Meyer (ed.), *British Popular Culture and the First World War*, Leiden: Brill, 217–236.

Pennell, C. and Sheehan, M. (2016) 'Official World War 1 memorial rituals could create generation uncritical of conflict', *Conversation UK*. Available from: https://theconversation.com/official-world-war-i-memorial-rituals-could-create-a-generation-uncritical-of-the-conflict-60384 (accessed 22 February 2017).

Price, J. (2005) 'Orphan heritage: Issues in managing the heritage of the Great War in northern France and Belgium', *Journal of Conflict Archaeology*, 1(1): 181–196.

Reynolds, D. (2013) *The Long Shadow: The Great War and the Twentieth Century*, London: Simon & Schuster.

Robertshaw, A. and Kenyon, D. (2008) *Digging the Trenches: The Archaeology of the Western Front*, Barnsley: Pen & Sword.

Roff, A. (2014) 'History calling', *Psychologies Magazine*, September: 50–53.

Saunders, N.J. (2001) 'Matter and memory in the landscapes of conflict: The Western Front 1914–1999', in B. Bender and M. Winer (eds.), *Contested Landscapes: Movement, Exile and Place*, Berg: Oxford, 37–53.

Scates, B. (2006) *Return to Gallipoli: Walking the Battlefields of the Great War*, Cambridge: Melbourne University Press.

Seaton, A.V. (1996) 'Guided by the dark: From thanatopsis to thanatourism', *International Journal of Heritage Studies*, 2(4): 234–244.

Seaton, A.V. (2000) '"Another weekend away looking for dead bodies . . .": Battlefield tourism on the Somme and in Flanders', *Tourism Recreation Research*, 25(3): 63–77.

Southern, K. (2016) '"Scum of the earth" and other things you're saying about vandals who targeted memorial to war dead', *Chronicle Live*, 31 May. Available from: www.chroniclelive.co.uk/news/north-east-news/scum-earth-things-youre-saying-11405195 (accessed 10 June 2016).

Switzer, C. and Graham, B. (2010) 'Ulster's love in letter'd gold': The Battle of the Somme and the Ulster memorial tower, 1918–1935', *Journal of Historical Geography*, 36(2): 183–193.

Vandaele, D. and Monballyu, M. (2008) 'Understanding WW1-related tourism in the Westhoek', in K. Lindroth and M. Voutilainen (eds.), *Competition in Tourism: Business and Destination Perspectives: Proceedings of the Travel and Tourism Research Association Europe–2008 Annual Conference, Helsinki*, Finland: Travel and Tourism Research Association-Europe, 362–371.

Van Emden, R. (2011) *The Quick and the Dead: Fallen Soldiers and Their Families in the Great War*, London: Bloomsbury.

Vanneste, D. and Foote, K. (2013) 'War, heritage, tourism, and the centenary of the Great War in Flanders and Belgium', in R. Butler and W. Suntikul (eds.), *Tourism and War*, London: Routledge, 254–272.

Wallis, J. (2015) '"Great-grandfather, what did you do in the Great War?": The phenomenon of conducting First World War family history', in B. Ziino (ed.), *Remembering the First World War*, Routledge: London, 21–38.

Walter, T. (1993) 'War grave pilgrimage', in I. Reader and T. Walter (eds.), *Pilgrimage in Popular Culture*, Basingstoke: Macmillan, 63–91.

Winter, C. (2009) 'Tourism, social memory and the Great War', *Annals of Tourism Research*, 36(4): 607–626.

Winter, C. (2011) 'Battlefield visitor motivations: Explorations in the Great War town of Ieper, Belgium', *International Journal of Tourism Research*, 13: 164–176.

Winter, J.M. (1995) *Sites of Memory, Sites of Mourning: The Great War in European Cultural History*, Cambridge: Cambridge University Press.

Winter, J.M. (2006) *Remembering War: The Great War between Memory and History in the Twentieth Century*, New Haven, CT: Yale University Press.

12 Witnessing the First World War in Britain

New spaces of remembrance

Ross Wilson

Introduction

This chapter examines the way in which the memory of the First World War in Britain has altered with the creation of new sites of remembrance to mark the centenary of the outbreak of the conflict from 2014. These commemorative spaces, both permanent and temporary, mark an alternative means of engaging with the conflict, formed as they are in conjunction with the existing commemorative landscape across the cities, towns and villages in Britain (King 1998). Erected in the decades after the conflict, these original constructions were designed to form a fixed location to remember the war dead. Whilst these sites of mourning remain significant within contemporary society, however, a shift in the spaces of remembrance has occurred in Britain. Since the 1990s, a wave of memorials connected to national and local concerns have engaged groups with the history of the conflict. Constructed by communities far removed from the direct experience of the war, these sites of memory are a means of forming a new commemoration of the conflict in the present day. Through a detailed assessment of the structure and content of these modern spaces of remembrance, this study examines the manner in which contemporary society in Britain is called to serve as witnesses to the war through these new locales (Wilson 2013). Using the recent emergence of a concern for memory studies within geography as a basis of analysis, this chapter develops approaches from non-representational theory to reveal how new sites of memory have structured commemoration as an act of observance (see Hoelscher and Alderman 2004; Thrift 2008). To perform the role of the witness, places moral, emotional and political obligations onto the individual since they are required to bear the burden of memory and to testify to its significance. The function of the witness within these memorial schemes will be examined, as these new spaces of remembrance demonstrate the way in which connections are made with the First World War a century after its outbreak.

The First World War: sites and locales of memory

The memory of the First World War occupies a peculiar place within contemporary British society. Even with the passing of the generation that experienced the conflict, to speak of the 'the trenches' or to mention the battlefields of the Somme, Gallipoli or Ypres is to evoke a poignant image of ruined landscapes, suffering soldiers, an expanse of headstones or the seemingly endless lists of names etched onto the memorials to the dead. Such a response could be assumed as inevitable considering the scale of loss in Britain, with over 700,000 deaths and over a million individuals wounded in varying degrees of severity (Winter 1985). In recent years, scholars have highlighted the constructed nature of this 'popular memory' of the war through an assessment of how architecture, literature, film and television have imagined and reimagined the conflict for successive generations since the Armistice in November 1918 (Bond 2002; Todman 2005; Hanna 2009). Within these analyses, the memory of the war is an 'invention of tradition'; a collection of responses born out of the way in which the war has been represented across the decades by authors, artists and politicians for varying agendas (Hobsbawm and Ranger 1983). The formation of the commemorative landscape in Britain and on the former battlefields during the 1920s and 1930s has been regarded as part of this process as it commandeered personal bereavement and established a sense of order for local and national authorities through dedicating the war dead to God, King and Empire (Bushaway 1992; Heffernan 1995; Gaffney 1998; Lloyd 1998).

With the publication of memoirs and novels by veterans of the conflict, such as Edmund Blunden (1928), Robert Graves (1929) and Siegfried Sassoon (1931), the late 1920s had challenged this image of glorious endeavour by presenting the war as a tragic failure. After the end of the Second World War, the image of the First World War as a failure of diplomacy and a dereliction of duty by the government was established through damning historical assessments that were formed in the shadow of the Cold War (see Taylor 1963). Theatre, literature, film and television programmes since the 1950s have also defined the conflict with a sense of pity, futility and waste to suit the politics, society and culture of postwar Britain. From *Oh! What a Lovely War* (1963), *For King and Country* (1965), *Blackadder Goes Forth* (1989) to *Regeneration* (Barker 1991) and *Birdsong* (Faulks 1992) the conflict is presented as an ill-planned and ill-judged travesty resulting in catastrophic loss (Williams 2009). These media representations have ensured the development of a distinct literary *topos*, where the war landscape of desolate, shell-scarred battlefields is inhabited by traumatised and disillusioned soldiers (Wilson 2014). Through these processes, the remembrance of the war in Britain has been formed as a contested site of memory with tensions between private and official commemoration as well as literary and historical representations (Wilson 2013).

Despite the varying ways in which the war has been regarded in Britain, what is a consistent feature in this battlefield of memory is the way in which notions of space have framed these discussions. Whether envisioned through the literary and media representation of the war or imagined through the memorial landscape in Britain or on the former battlefields, this is an act of commemoration that is structured by the places of remembrance. This distinction is significant as it enables the recognition of how these spaces are not inert but rather used by groups as a means of defining themselves; these places of remembrance are enacted and employed (after Harvey 2001). The notion of utility is significant in this respect; for whilst assessments of the 'popular memory' of the war regard the way in which emotion, sentimentality and historical reductionism through various media have shaped commemoration, these studies have tended to neglect the way in which remembrance functions for individuals, communities and wider society (after Wertsch 2002). Rather than solely manipulated by the representations of the conflict through art, architecture, literature, film or television, the remembrance of the war is an active pursuit undertaken by present-day communities for a purpose. Through these sites of memory, whether figurative or literary, the commemoration of the conflict stakes a claim to identity, power and authority. The memory of the First World War in Britain, therefore, is adapted and altered by those who mark it and formed through notions of space and place.

By focusing on location, value and utility within the remembrance of the conflict in Britain, a number of scholars within the field of cultural and historical geography have analysed how the civic and national memorial landscape was formed in the wake of the war's conclusion (Graham et al. 2000; Graham and Shirlow 2002; Johnson 2003). This work has influenced scholarship within tourism studies which assesses the contemporary relevance of these spaces for society through examining practices of commemoration such as battlefield tours (Iles 2006, 2008; Winter 2009a, 2009b, 2010). However, whilst the assessment of the history and heritage of the original memorials and monuments built to commemorate the war dead is an established field, the study of the new spaces to mark the conflict which have been built in recent years has received relatively little examination (see Goebel 2004). This absence neglects a developing and significant field as these locales of remembrance have been built since the early 1990s and have served as commemorative spaces for a generation removed from the direct experience of the war. These memorials represent a shift in the form and function of the remembrance of the war as emotional, moral and political identities are created within these sites for modern concerns. Whilst influenced by previous commemorative practices, it is through these new spaces that contemporary society bears witness to the events of 1914–1918.

Space, place and memory

The role of geography within the wider study of remembrance, commemoration and heritage has been enhanced in recent years through the inclusion of a critical and performative perspective (Harvey 2001; Garde-Hansen and Jones 2012). These analyses have defined how particular uses of space and place can serve as a means by which a past is remembered and reimagined to form identities, establish authority and counter-hegemonic structures of power. As such, the connections between space, place and memory have been explored by scholars who have established the role of sites of remembrance within contemporary political, social and cultural discourses (Hoelscher and Alderman 2004; Jones 2011). These assessments have tended to concentrate upon the places that evoke the traumatic histories of the twentieth century with a focus on the interaction of individuals with these places of sorrow, bereavement and loss (see Winter 1992; Dawson 2005; Johnson 2012). Spaces of remembrance are examined as communicative locales, where ideas, values and agendas are formed and challenged through an engagement with the physical environment. It is in this regard that the develop-ment of non-representational theory within geography has been forwarded as a means by which this performance of memory can be analysed. This approach is defined by a series of concerns rather than a rigid methodology that seeks to place the performance, engagement and agency at the centre of assessment (Lorimer 2005). The principles of non-representational theory emerged with a concern that the practices of life had been obscured with a focus on representation rather than engagement (Thrift 2003, 2008). Emotional and affective responses to objects, spaces, places and landscapes can thereby be regarded as central to understanding the use of sites (Thrift 2008).

The limitations of non-representational theory are apparent in its assessment of affect that focuses on distinct points in time and space. This concentration obscures the manner in which performances of memory in places are composed of multiple layers of meaning, formed through imagined associations and his-torical experience (see Hill 2013). This is especially the case with regard to the remembrance of the First World War in Britain; the landscape of commemora-tion is a palimpsest, defined by successive generations who have engaged with the memory of the war (after Harvey 2013). The new sites of remembrance that have emerged within Britain over the last three decades not only engage with the history of the conflict itself, but also the legacy of its commemoration within British society from its immediate aftermath to the centenary of its outbreak. Whilst the performances and engagements with these new spaces of memory can be examined, this must be conducted alongside an assessment of how such practices are situated within a wider context and tradition of the commemoration of the First World War. The use of non-representational theory to assess the emo-tive and affective engagement with the First World War provides a lens through which to understand how individuals, groups and communities have formed their particular practices of remembrance through the tangible and intangible heritage of the conflict (after Wertsch 2002). These modern sites of memory, therefore, can be regarded as enabling a 'witness' perspective on the past and the pres-ent. Such an engagement reflects both the history of commemoration and the contemporary performance of memory that is enacted within these spaces (after Dewsbury 2003).

The significance of the 'witness' is detailed within a legal and religious context in Western culture, which requires acknowledgement, observance and testimony from the individual (see Derrida 2005: 75–79). As a first-hand observer or as a bearer of knowledge, the witness serves to ensure the remembrance of particular events (see Douglass and Vogler 2003). Such acts of witnessing are not neutral records, but rather performances that enable the creation of memory that acknowl-edges the past and addresses the present (LaCapra 1994; Apel 2002). Significantly, this witnessing is bound by and defined by notions of space, place and purpose:

> Places are witnesses, locales of memory that we mark out or that simply are
> there waiting, traces that serve to remind us of those things that need remem-
> bering, for which there is a duty of one kind or another for us to bear witness.
>
> (Booth 2006: 111)

The act of bearing witness to the First World War in Britain is performed through
these new places of commemoration. This observance denotes emotional, moral,
political and social duties onto the individual as they are made to bear the burden of
memory and, by doing so, testify to its significance in the present. The new geography of remembrance is formed in association with the original sites of memory but
it marks an alternative engagement with the legacy of the war. Modern witnesses to
the conflict claim association with the suffering and trauma of the war as a means
of asserting and challenging notions of identity and structures of power within
contemporary Britain. In these new spaces of memory, the effects of the conflict are
witnessed and testimonies are formed to make this past war appear present.

Emotional spaces of remembrance

Over the last three decades, cities and towns in Britain began to erect new memorials to soldiers of the First World War. This commemoration was not born out of
an absence of individual and collective commemorative schemes. Indeed, many
of these areas already possessed memorials which were built during the interwar
period and which have been central in commemorating the dead of the First World
War and from subsequent conflicts throughout the twentieth century.[1] New memorials, therefore, formed alternative spaces of remembrance for a war that had
already been marked within the commemorative landscape. What is distinct about
these sites is the emotional engagement that is required from the witness. Where
the original commemorative designs were constructed to affirm the values of the
nation and were used by the directly bereaved, these new structures have been
formed as an affective space for contemporary audiences. The places chosen for
these modern sites of mourning are significant, in that many of these memorials
are placed in areas of emotional intensity for the history of the conflict in Britain.

   This is perhaps best illustrated with the siting of centenary memorials in areas
of departure, where recruits would have paraded before being relocated to training
or where soldiers would have been separated from friends or family as they were
deployed or returned to the front. This emotional space is intimately linked to a
sense of place and locality for witnesses to bear testimony to the effect of the war.
For example, a memorial garden and plaque were unveiled at Letchworth Railway
Station (Hertfordshire) in August 2014, which accentuated the sense of loss and
trauma, dedicated as it was, according to the local newspaper, to 'men who left
the town 100 years ago, never to return' (see Scott 2014). In this memorial space,
alongside the display of poppies, a photograph of soldiers waiting on the railway
platform is accompanied by a panel detailing where local servicemen lived within
the town. Similarly, in November 2014, wooden plaques were unveiled at Paragon Street Station in Hull (East Yorkshire) that listed the names of two thousand

men from the city who were killed during the war. The placing of the plaques by the entrance of the building was regarded as necessary to intensify the emotional connection to the history as the place where soldiers 'travelled from the station to fight abroad, but never returned' (Roberts 2014). The same mode of memorialisation can be observed at Liverpool's Lime Street Station (Merseyside) in August 2014, where two bronze plaques were unveiled to commemorate the service of the 'Liverpool Pals' at their final point of departure from the city. Whilst in 2012, a slate plaque was raised at Preston Railway Station (Lancashire) to mark the site where the 'Preston Pals' left for the battlefields.

These modern memorials share points of connection with the commemorative schemes that were established in the aftermath of the war, as plaques or tablets at railway stations or nearby were a common aspect of remembrance during the 1920s. These sites were used as their central place within the urban landscape ensured a public engagement with the commemoration of the war and because of the importance of the railways in both the recruitment and deployment of servicemen (see Moriarty 2003: 41). This can be seen in the memorial unveiled in 1922 at Stoke Railway Station, where the names of over 100 employees of the North Staffordshire Railway Company were commemorated on plaques at the entrance to the platforms. Similarly, a plaque was unveiled at Perth General Station in 1921 to the memory of 11 members of staff who lost their lives during the war. Speaking at its dedication, John Stewart-Murray, eighth Duke of Atholl (1871–1942), remarked, 'The splendid spirit of sacrifice, courage, endurance, and loyalty of the men' was a 'precious heritage' (Anon 1921). These memorials, fusing employment, nationalism and local identity, provided a 'corporate' commemorative space to observe the service of individuals (after Gough 2004). Whilst important for the bereaved, these memorials marked the wartime dedication to victory and their public utility was based on their ability to enable performances of memory that signify the sacrifice but enable an orientation towards the future. The memorial on Station Road, Cambridge, illustrates this effect as the sculpture of a uniformed soldier looking back to the railway station but marching towards the city is entitled 'The Homecoming' (Inglis 1992) (Figure 12.1).

A contrast is formed between the past and present spaces of remembrance where the former evokes sacrifice whilst the latter enables contemporary society to witness the trauma of the war. This particular distinction can be observed at the 'Soldier's Gate', a memorial in Manchester's Victoria Station, which was part of the original commemorative scheme that was erected in 1922. The gate was the site from which troops would have arrived and departed by railway for the front in France and Belgium and a plaque marks the point:

TO THE MEMORY OF THE MANY THOUSANDS OF MEN WHO PASSED THROUGH THIS DOOR TO THE GREAT WAR 1914–1919 AND OF THOSE WHO DID NOT RETURN

Nevertheless, with the redevelopment of Victoria Station in Manchester in November 2015, the 'Soldiers' Gate' was preserved but the space was further

Figure 12.1 Cambridge War Memorial.

developed with a displayed map detailing the location of all the major cemeteries in France and Flanders. This new memorial was also explained with the addition of a plaque placed below the original but in the same design that outlined how the map detailed the loss of life on the Western Front. In the new geography of remembrance, the enactment of trauma defines the role of the witness (after Thrift 2003: 2019–2020). Whilst the original memorials provide a space to consider the service and sacrifice of the nation, modern memorials evoke the grief and bereavement for contemporary society. In effect, this serves to assert the remembrance of the war as a tragic waste through a 'mimetic mechanism' as the sense of loss from the past is imitated in the present (Thrift 2008: 232). Rather than a simple mimicry of the past, this emotional response is a product of active engagement with the legacy of the war (after Taussig 1993). Contemporary witnesses use the new memorials placed in railway stations over the past two decades to emphasise the sentimental and the sorrowful. This can be noted in the 2008 renovation of the Dingwall Railway Station in Ross and Cromarty (Scotland), where a plaque was erected by the Railway Heritage Committee that detailed the scale of the station's wartime function in providing refreshment:

THIS RAILWAY STATION WAS USED AS A TEA STALL FOR SOL-DIERS AND SAILORS FROM 20th SEPTEMBER 1915, UNTIL 12th APRIL 1919 IN CONNECTION WITH THE ROSS AND CROMARTY COUNTY BRANCH RED CROSS SOCIETY DURING WHICH PERIOD 134,864 MEN WERE SUPPLIED WITH TEA.

The creation of new spaces of grief and trauma to mark the war represents a means by which contemporary society forms a relationship to this past in the present (after Dewsbury 2003). From the 1990s as a discernible shift towards focusing on the pain and sorrow of the conflict can be noted as a means by which the war is made relevant for individuals, groups and communities. For example, the Western Front Association, which formed in 1980 to further the study of the conflict, sponsored a plaque in 1998 on the platform of London's Victoria Railway Station that commemorates where the body of the 'Unknown Warrior' was placed before burial:

THE BODY OF THE BRITISH UNKNOWN WARRIOR ARRIVED AT PLATFORM 8 AT 8.32PM ON THE 10TH NOVEMBER 1920 AND LAY HERE OVERNIGHT BEFORE INTERMENT AT WESTMINSTER ABBEY ON 11TH NOVEMBER 1920.

To bear witness to the war in the modern era is, therefore, to regard the trauma, loss and sorrow of the conflict. Spaces of remembrance have been constructed in the last two decades that connect contemporary society to the events of 1914–1918 on the basis of emotional engagement (after Dewsbury 2003). To be a witness to the war is to regard its devastating consequences for individuals and communities in the past. These spaces of remembrance enable the testimony of the war's remembrance as a shocking and disturbing event. This is heightened by their siting at railway stations, which emphasises the sense of separation and bereavement. This mode of witnessing provides a means of continually engaging with this history for contemporary concerns; the emotional attachments formed at this site can act as an affirmation of collective effort, a demonstration of the waste of the war or as an indication of regional identity. Regardless of the effect to which these spaces are used for it is their mode of affect that marks them out as a modern memorial to the war. To witness the emotion of the conflict within these sites is to testify to the relevance of the past in the present.

Moral spaces of remembrance

Within the last few decades, a new wave of memorial spaces have been constructed in Britain that form 'moral witnesses' to the conflict fought at the outset of the twentieth century. This particular commemoration of the First World War is distinctive in the way it requires a performance of memory where the individual is obliged to bear the burden of remembrance. It marks a distinct departure from the memorials that were dedicated in the wake of the war's denouement that tended to evoke a collective sacrifice (King 1998). In this manner, moral spaces of remembrance are a specifically modern phenomenon. Margalit (2002) defined a 'moral witness' as a figure who observes suffering, pain or trauma and testifies as to its occurrence and its significance. Rather than being an accurate recorder of historical events, the perspective of the 'moral witness' can be limited in scope or analysis but that does not lessen the importance of the role (Winter 2006: 242). What is important about the 'moral witness' is the expectation that the testimony, the performance of remembrance, is used to rectify, reform or to call for justice. This process can be assessed within the new memorials to the war of 1914–1918, as these spaces of commemoration serve to exercise moral perspectives within contemporary society.

This can be observed with the new memorials and the adjustments to existing memorials that mark the deaths of those soldiers executed by the British Army during the war. With the success of the campaign in 2006 to provide pardons for over 300 individuals who were court martialled, sentenced to death and shot by firing squad, a demand for recognition of these cases emerged. This led to an alteration to many post-war memorials in towns and cities where the names of those executed are placed onto tablets, plaques and monuments. In 2014, the name of Private William Jones was placed on the Glynneath Memorial (Glamorgan, Wales), which had been built in the 1920s to honour the sacrifice of men from the region in the war. Private William Jones had absconded from military service in France and returned to his family home, where he was convinced to turn himself over the authorities. He was executed for desertion in October 1917, but the inclusion of his name on the war memorial has served to transform the site of remembrance for the local area. The historian, Robert King, who campaigned for inclusion of Private William Jones stated at the outset of the campaign:

It was futile and it was for the sake of example, and only working class men were shot. It is a dark part of military history.

(quoted in Anon 2013)

The creation of 'moral spaces of remembrance' has been driven by the movement to have soldiers who were 'shot at dawn' to have their convictions overturned. In 1998, nearly 40 names of soldiers who had been executed for crimes including cowardice and desertion were added to the book of remembrance in the Scottish National War Memorial within Edinburgh Castle. Although this addition was conducted, according to reports, without 'fuss or publicity', it did mark an alteration in this site of mourning; where once the memorial housed the names of those who died in the war it now honoured the memory of all those regarded as 'victims' of the conflict (Anon 1998). In 2001, this perception of victimhood and the formation of moral spaces of remembrance was confirmed with the dedication of the 'Shot at Dawn' memorial in the National Arboretum in Staffordshire. This statue, modelled on Private Herbert Burden who was killed by firing squad at the age of 17, is a blindfolded and manacled individual who stands in front of an array of large wooden stakes that bear the names of those executed. The subject of the status of those British soldiers was resolved in 2006, with Parliament granting pardons to all individuals executed by the British Army. Consequently, the space in the National Arboretum requires visitors to become 'moral witnesses' to the conflict, bearing the burden of memory and compelled to observe the traumatic effect of the war on the lives of individuals and families. Indeed, the media reporting of the memorial at its unveiling and in its immediate aftermath focused upon this capacity of the memorial to ensure recognition for the past in the present:

> Britain yesterday proudly saluted the 306 servicemen shamefully shot by their own country during the First World War. A 10ft-high statue of a teenage soldier blindfolded and tied to a stake, entitled simply Shot at Dawn, was unveiled in a moving ceremony.
>
> (Roberts 2001)

> The victims were perceived as the "low-bred dregs of humanity, degenerate and worthless". They were not murderers, rapists or criminals; they were, in most cases, men whose crime was to fear death or suffer from psychiatric injury caused by the hell that was the First World War.
>
> (McBeth 2001)

The memorial at the National Arboretum also encouraged further alterations at existing sites where the names of executed soldiers were placed upon memorials built in the 1920s that did not bear the service of these individuals. For example, in February 2007 the 1923 memorial at Wealdstone Clock Tower (London) was altered to include the names of Private Harry Farr and Private James Swain who were killed by firing squad in 1917. The rededication of sites or remembrance to include those regarded as 'victims' of the war emphasises the formation of moral spaces of remembrance within the last two decades. As the anniversaries of the war have been marked and the veterans of the conflict have passed away, there has emerged a particular focus on the way in which witnesses can testify to the trauma and tragedy of the conflict. This can be seen in the 1994 dedication of the Conscientious Objectors Stone in Tavistock Square (London), a large granite rock that bears a plaque with the inscription:

> To all those who have established and are maintaining the right to refuse to kill. Their foresight and courage give us hope.

Whilst dedicated to a universal cause, the memorial was erected in the eightieth anniversary year of the outbreak of the conflict and the Peace Pledge Union, who coordinated the project, drew upon the specific legacy of those who had refused military service during the First World War. Indeed, a ceremony to mark the lives of those conscientious objectors who rejected military service during the conflict was held at this site on May 15 2014. Through this performance, the tangible and intangible effects of remembrance provided a space through which individuals could serve as moral witnesses to the war. A similar site was completed in 2002 at Richmond Castle (North Yorkshire), where a memorial garden stands in the grounds to commemorate the 16 individuals who were held in the castle prison during the war. Whilst initially attracting some degree of criticism as to the function of remembering these individuals, the site has become a central part of remembrance activities.

Such is the nature of this mode of witnessing that the marking of the centenary of the outbreak of the conflict was accompanied by calls for other permanent memorials to the conscientious objectors. In 2015, Edinburgh residents were successfully petitioned by the city council on behalf of religious and anti-war groups to gather support for a memorial to those who refused to fight for their country during the First World War. Residents were asked to consider the necessity of a space for remembering on the basis of moral need to bear witness to this history:

> With the Centenary of the First World War there is a feeling that there should be a memorial in Scotland's capital city to conscientious objectors and opponents of wars which would henceforth provide a public focus for those who wish to gather to remember all those, past or present, refusing to participate in or opposing wars.
>
> (Edinburgh City Council 2015).

The use of memorials to create spaces where the voices of those who have been suppressed or obscured is also present in the commemoration of soldiers of African and African Caribbean heritage who served in the British Army during the conflict. A memorial to the African British soldier and professional footballer, Walter Tull (1888–1918), was unveiled outside Tull's former team Northampton Town's Sixfields Stadium in 1999. Inscribed with a message that emphasises Tull's role in 'ridiculing the barriers of ignorance', the memorial site asserts the need for moral remembrance to understand the experience of those in the past and the necessity of acknowledgement in the modern era. Similarly, the 2014 dedication of the African and Caribbean War Memorial outside the Black Cultural Archives in Brixton (London) formed a space through which a recognition of injustice could be observed. The two obelisks that form the memorial are inscribed with the names of the African and Caribbean regiments that served in the British Army during 1914–1918 and 1939–1945. Whilst dedicated to both conflicts, it is the anniversary of the outbreak of the First World War that was behind the construction of the site. The commemoration also bears the invocation 'Remembering the Forgotten', emphasising the need to address an absence in the initial commemorative schemes. Indeed, Aloun Ndombet-Assamba, High Commissioner for Jamaica, speaking at the unveiling ceremony stated,

> This year, as the one hundredth anniversary of the start of the First World War is observed, it is fair to say that the role played by the Caribbean, Africa, India and other parts of Asia is still not widely known by many in Britain. The Caribbean and Africa were profoundly affected by the war as manpower, materials, and funds were sent by them to the aid the war effort to protect Britain and Europe.
>
> (Anon 2014)

Over the last three decades, the remembrance of the war in Britain has been altered by these moral spaces of remembrance where the injustices of the past or the inequality of history are witnessed and brought to bear on the present. These new sites of memory have ensured that the commemoration of the conflict in the modern era is framed as an ethical concern. A sense of emotional attachment and place is formed with the past through this witnessing (after Dewsbury 2003).

Socio-political spaces of remembrance

The memorials built in the villages, towns and cities of Britain after the First World War have been examined by scholars as political statements which ensured stability within a turbulent era (Connelly 2002). These local memorials echoed the themes and designs of the central memorial schemes; the Tomb of the Unknown Warrior in Westminster Abbey, the Cenotaph in Whitehall and the memorial landscape on the former battlefields. The inauguration of the two minutes silence for Armistice Day (1919) also brought comfort to the bereaved and sustained an image of the war as a noble sacrifice (after Mosse 1990). This process fashioned spaces and practices of political remembrance which committed witnesses to the ideals and values of the state (after Bushaway 1992; Heffernan 1995). With the advent of the centenary of the declaration of war, new sites of political memory have been developed which serve to inculcate the perspective of the collective sacrifice. These sites echo the imagery and symbolism of their predecessors to create spaces of witnessing which emphasise a common purpose. These new places of remembrance have appeared as a specific response to the anniversary of the war's outbreak and reflect the difficulties of remembering a war whose meaning is highly contested within Britain (see Bond 2002; Todman 2005). Such difficulties were present in the development of commemorative schemes in the lead-up to the centenary year of 2014. Prime Minister David Cameron defined the official themes of remembrance at the Imperial War Museum in November 2012:

> Our duty towards these commemorations is clear: to honour those who served, to remember those who died, and to ensure that the lessons learnt live with us forever.
>
> (Cameron 2012)

This structured remembrance is best demonstrated in the orchestration of the 'Lights Out' programme on August 4 2014 where households and businesses were encouraged to turn off any lights for an hour except for one from 10pm in a 'shared moment of reflection' (14–18 Now 2014). Inspired by the comments of the wartime Foreign Secretary Sir Edward Grey (1862–1933) that 'the lamps are going out all over Europe', and organised by an arts initiative funded by the Imperial War Museum, the Heritage Lottery Fund and government offices, this action served to create new commemorative spaces across Britain that focused on a sense of collective endeavour. Individuals, groups and communities were able to bear witness to the sacrifices made to the nation.

These particular sites of memory that focus upon establishing common values and ideals are in contrast to those sites that address moral concerns. This can be observed in the large-scale installation of ceramic poppies around the Tower of London in the autumn of 2014. Entitled 'Blood Swept Lands and Seas of Red', this artwork was assembled from 888,246 individual flowers to commemorate each of the lives lost by British or Dominion servicemen during the war (Figure 12.2). The formation of the memorial over several weeks drew public and media attention and it became a collective site of mourning on its completion on November 11. Responses to the piece reflected upon the shared sacrifice implied by the vast nature of the commemorative scheme:

> Seeping from the walls like a wound that never healed, a growing sea of blood-red poppies has transformed the Tower of London into the nation's rallying point of remembrance.
>
> (McPhee 2014)

> They were the flower of British youth and they gave their lives for our freedom in the Great War a century ago. So it seems the perfect tribute that the Tower of London, a timeless symbol of our nation, should be weeping for them today, on Remembrance Sunday.
>
> (Bletchly 2014)

With millions of visitors recorded at the site, the memorial created a space where a shared sense of loss could be reflected upon. Significantly, this particular place

Figure 12.2 'Blood Swept Lands and Seas of Red', Tower of London, November 2014.

of remembrance was not offered to emphasise a moral cause or perspective in the present but as a means of establishing collective witnesses to the service of the nation. Whilst certainly an emotive subject, this artwork nevertheless served as a means of generating public recognition as to the scale of sacrifice during the First World War. This process can also be observed in the unveiling of a large, steel Memorial Arch in Folkestone (Kent) on the site where soldiers boarded ships for transportation to France. The structure was dedicated in August 2014, replacing a temporary arch that had been sited there in the 1930s. Speaking at the dedication ceremony, the local member of Parliament, Damian Collins, stated the importance of the site for the nation and its role in preserving the shared sense of service to the nation (Collins 2014). Indeed, a plaque on the arch dedicates the structure to this cause:

COMMEMORATES THE MILLIONS OF MEN AND WOMEN WHO PASSED THROUGH FOLKESTONE IN THE SERVICE OF THEIR COUNTRY DURING THE FIRST WORLD WAR

These space of remembrance form a means by which individuals and groups can congregate and form new attachments in a manner not dissimilar to the 'communities of mourning' formed through the commemorative sites built in the aftermath of the First World War (see Winter 1992). These new places of memory structure this act of witnessing as a social act but do not enable a sense of testimony where the witness reflects on the emotional or moral effect on the present. The purpose of this witnessing is proscribed, the values and meanings of these sites are already defined, and the witness is required to observe and acknowledge. This is most clearly observed with the government-funded scheme to provide commemorative paving stones to regional authorities to remember recipients of the Victoria Cross (VC), Britain's highest military award, who were born in the local area. Details provided by the Department of Communities and Local Government (DCLG) and distributed in May 2014 required these stones to be placed in particular locales:

> It is hoped that the stones will be sited somewhere that would have had reso-nance with the VC recipient, such as outside a house that they lived in or near an old school, although this may not always be possible. Most importantly, the stones should be part of the community and sited in a position where they will be visible to members of the public.
>
> (DCLG 2014)

In effect, this forms a space where individuals and communities can attend and be informed of what to witness in their remembrance rather than utilise that testimony to effect the present. This is not to deny the emotional attachment to these places of memory or their significance to the communities in which they are sited. Rather, it demonstrates how these effects are structured by the testimony that has already been defined. As opposed to the moral or emotional spaces of remembrance, the socio-political places of memory formed for the centenary act to reduce the role of the witness to a passive bearer of information. This neutralises or constrains the meaning of the war for contemporary society and ensures the spaces of com-memoration control the act of witnessing. To observe the remembrance of the First World War through these spaces is to mark a common experience but it obscures an engagement with the effects of the conflict in the past and in the present.

Conclusions

In the formation of new sites of memory of the First World War in Britain, the act of witnessing is crucial. Since the formation of memorial schemes after the conclusion of the conflict, sites of memory have orientated individuals, groups and communities in particular ways in relation to the events of 1914–1918. These places of memory have structured the remembrance of the war and the develop-ment of new memorials over the past three decades has continued this process. From the eightieth anniversary of the onset of war in 1994, new sites have been built which seek to engage a society that is far removed from the experience of the conflict. Rather than assessing these alternative spaces of the war for the con-temporary performances and practices of memory to form a synchronic mode of analysis, an alternative model of 'the witness' can be used to understand the engagement of the past and present forms of remembrance. This emphasises the alterations and adaptations in which the war is remembered in the present day through these sites of memory. New sites of memory constitute a dialogue with the original spaces of remembrance and it is through that interaction that modern places of commemoration can be examined. The contemporary sites demonstrate continuity and change in the remembrance of the conflict as new spaces are built that enable other modes of witnessing to take place. What has emerged in the recent formation of new places of commemoration is a means of testifying to the effects of the war; memorials provide a means of actively engaging in the history of the conflict to emphasise the effect of the war in the past and the present.

Note

1 Remembrance Day in Britain is held on November 11. This day commemorates the sign-ing of the armistice in 1918 and was used subsequently as an anniversary to remember the war dead. After the Second World War, November 11 has become the day in which all of those who have been killed in the nation's wars are commemorated.

Bibliography

14–18 Now (2014) *Lights Out.* [Online]. Available from: www.1418now.org.uk/commissions/lightsout/ (accessed 5 August 2014).

Anon (1921) 'Perth station War Memorial', *Dundee Courier*, 17 January, 2.

Anon (1998) 'Firing squad victims added to memorial', *The Herald*, 14 July, 8.

Anon (2013) 'Add the deserters', *Wales Online*. [Online]. Available from: www.walesonline.co.uk/news/local-news/add-the-deserters-2323588 (accessed 4 December 2014).

Anon (2014) 'Memorial to "overlooked" African and Caribbean war veterans', *The Voice*. [Online]. Available from: www.voice-online.co.uk/article/memorial-overlooked-african-and-caribbean-war-veterans (accessed 12 November 2014).

Apel, D. (2002) *Memory Effects: The Holocaust and the Art of Secondary Witnessing*, New Brunswick, NJ: Rutgers University Press.

Bletchly, R. (2014) 'The fallen fleetingly in flower', *The People*, 9 November, 3.

Blunden, E. (1928) *Undertones of War*, London: Cobden-Sanderson.

Bond, B. (2002) *The Unquiet Western Front: Britain's Role in Literature and History*, Cambridge: Cambridge University Press.

Booth, W.J. (2006) *Communities of Memory: On Witness, Identity, and Justice*, Ithaca: Cornell University Press.

Bushaway, B. (1992) 'Name upon name: The Great War and remembrance', in R. Porter (ed.), *Myths of the English*, Cambridge: Polity Press, 136–167.

Cameron, D. (2012) 'Speech at Imperial War Museum on First World War centenary plans'. [Online]. Available from: www.gov.uk/government/speeches/speech-at-imperial-war-museum-on-first-world-war-centenary-plans (accessed 10 November 2013).

Collins, D. (2014) 'War centenary events a great achievement', *Folkestone Herald*, 7 August. [Online]. Available from: http://www.damiancollins.com/blog/step-short-event-a-poignant-commemoration/ (accessed 22 November 2014).

Connelly, M. (2002) *The Great War, Memory and Ritual: Commemoration in the City and East London, 1916–1939*, London: Boydell.

Dawson, G. (2005) 'Trauma, place and the politics of memory: Bloody Sunday, Derry, 1972–2004', *History Workshop Journal*, 59(1): 151–178.

DCLG (2014) 'Victoria Cross commemorative paving stones guidelines for Councils'. [Online]. Available from: www.gov.uk/government/uploads/system/uploads/attachment_ data/file/319813/Guidelines_for_Councils_220514.pdf (accessed 5 June 2015).

Derrida, J. (2005) *Sovereignties in Question: The Poetics of Paul Celan*, New York: Fordham University Press.

Dewsbury, J.-D. (2003) 'Witnessing space: "Knowledge without contemplation", *Environment and Planning A*, 35(11): 1907–1932.

Douglass, A. and Vogler, T.A. (eds.) (2003) *Witness and Memory: The Discourse of Trauma*, London and New York: Routledge.

Edinburgh City Council (2015) 'Edinburgh conscientious objectors memorial petition'. [Online]. Available from: www.edinburgh.gov.uk/site/xfp/scripts/xforms_form. php?formID=321&language=en (accessed 9 October 2015).

Gaffney, A. (1998) *Aftermath: Remembering the Great War in Wales*, Cardiff: University of Wales Press.

Garde-Hansen, J and Jones, O (eds.) (2012) *Geography and Memory*. Basingstoke: Palgrave Macmillan.

Goebel, S. (2004) 'Re-membered and re-mobilized: The "sleeping dead" in interwar Germany and Britain', *Journal of Contemporary History*, 39(4): 487–501.

Gough, P. (2004) 'Corporations and commemoration: First World War remembrance, Lloyds TSB and the National Memorial Arboretum', *International Journal of Heritage Studies*, 10(5): 435–455.

Graham, B., Ashworth, G.J. and Tunbridge, J.E. (2000) *A Geography of Heritage: Power, Culture and Economy*, London: Arnold.

Graham, B. and Shirlow, P. (2002) 'The Battle of the Somme in Ulster memory and identity', *Political Geography*, 21: 881–904.

Graves, R. (1929) *Goodbye to All That*, London: Jonathan Cape.

Hanna, E. (2009) *The Great War on the Small Screen: Representing the First World War in Contemporary Britain*, Edinburgh: Edinburgh University Press.

Harvey, D. (2001) 'Heritage pasts and heritage presents: Temporality, meaning and the scope of heritage studies', *International Journal of Heritage Studies*, 7(4): 319–338.

Harvey, D. (2013) 'Emerging landscapes of heritage', in P. Howard, I. Thompson and E. Waterton (eds.), *The Routledge Companion to Landscape Studies*, London: Routledge, 152–165.

Heffernan, M. (1995) 'For ever England: The Western Front and the politics of remembrance in Britain', *Ecumene*, 2: 293–324.

Hill, L. (2013) 'Archaeologies and geographies of the post-industrial past: Landscape, memory and the spectral', *Cultural Geographies*, 20(3): 379–396.

Hobsbawm, E. and Ranger, T. (1983) *The Invention of Tradition*, Cambridge: Cambridge University Press.

Hoelscher, S. and Alderman, D.H. (2004) 'Memory and place: Geographies of a critical relationship', *Social & Cultural Geography*, 5(3): 347–355.

Iles, J. (2006) 'Recalling the ghosts of war: Performing tourism on the battlefields of the Western Front', *Text and Performance Quarterly*, 26(2): 162–180.

Iles, J. (2008) 'Encounters in the fields: Tourism to the battlefields of the Western Front', *Journal of Tourism and Cultural Change*, 6(2): 138–154.

Inglis, K.S. (1992) 'The homecoming: The war memorial movement in Cambridge, England', *Journal of Contemporary History*, 27(4): 583–605.

Johnson, N.C. (2003) *Ireland, the Great War and the Geography of Remembrance*, Cambridge: Cambridge University Press.

Johnson, N.C. (2012) 'The contours of memory in post-conflict societies: Enacting public remembrance of the bomb in Omagh, Northern Ireland', *Cultural Geographies*, 19(2): 237–258.

Jones, O. (2011) 'Geography, memory and non-representational geographies', *Geography Compass*, 5(12): 875–885.

King, A. (1998) *Memorials of the Great War in Britain: The Symbolism and Politics of Remembrance*, Oxford: Berg.

LaCapra, D. (1994) *Representing the Holocaust: History, Theory, Trauma*. Ithaca: Cornell University Press.

Lloyd, D. (1998) *Battlefield Tourism: Pilgrimage and the Commemoration of the Great War in Britain, Australia, and Canada, 1919–1939*, Oxford: Berg.

Lorimer, H. (2005) 'Cultural geography: The busyness of being "more-than-representational"', *Progress in Human Geography*, 29(1): 83–94.

Margalit, A. (2002) *The Ethics of Memory*, Cambridge, MA, Harvard University Press.

McBeth, J. (2001) 'Shot at dawn', *The Scotsman*, 9 November. [Online]. Available from: http://www.scotsman.com/news/shot-at-dawn-1-583451 (accessed 22 November 2014).

McPhee, R. (2014) 'We treat each flower with reverence because each one really does represent a life', *The Mirror*, 11 August. [Online]. Available from: http://www.mirror.co.uk/news/uk-news/inside-ceramic-poppy-factory-workers-4034309 (accessed 22 November 2014).

Moriarty, C. (2003) 'Though in a picture only: portrait photography and the commemoration of the First World War', in G. Braybon (ed.), *Evidence, History and the Great War: historians and the impact of 1914-18*, Oxford: Berg, 30–47.

Mosse, G. (1990) *Fallen Soldiers: Reshaping the Memory of the World Wars*, Oxford: Oxford University Press.

Roberts, A. (2014) 'Paragon station memorial to Hull's First World War dead to be unveiled on remembrance Sunday', *Hull Daily Mail*, 11 October. [Online]. Available from: http://www.hulldailymail.co.uk/paragon-station-memorial-hull-s-world-war-dead/story-23098000-detail/story.html (accessed 18 November 2014).

Roberts, B. (2001) 'Soldiers WE shot at dawn', *The Mirror*, 22 June, 4.

Sassoon, S. (1931) *Memoirs of an Infantry Officer*, London: Faber and Faber.

Scott, J. (2014) 'First World War memorial and plaques unveiled at Letchworth railway station', *The Comet*, 4 August. [Online]. Available from: http://www.thecomet.net/news/first-world-war-memorial-and-plaques-unveiled-at-letchworth-railway-station-1-3713159 (accessed 18 November 2014).

Taussig, M. (1993) *Mimesis and Alterity: A Particular History of the Senses*, New York: Routledge.

Taylor, A.J.P. (1963) The First World War: An Illustrated History, London: Penguin.

Thrift, N. (2000) 'Afterwords', *Environment and Planning D: Society and Space*, 18: 213–255.

Thrift, N. (2003) 'Performance and . . . .', *Environment and Planning A*, 35: 2019–2024.

Thrift, N. (2008) *Non-Representational Theory: Space, Politics, Affect*, Abingdon and New York: Routledge.

Todman, D. (2005) *The Great War: Myth and Memory*, London: Hambledon.

Wertsch, J.V. (2002) *Voices of Collective Remembering*, Cambridge: Cambridge University Press.

Williams, D. (2009) *Media, Memory*, and the *First World War*, Montreal and Kingston: McGill-Queen's University Press.

Wilson, R.J. (2013) *Cultural Heritage of the Great War*, Farnham: Ashgate.

Wilson, R.J. (2014) 'It still goes on: Trauma and the memory of the First World War', in M. Sokolowska-Paryz and M. Löeschnigg (eds.), *Great War in Post-Memory Literature, Drama and Film*, Berlin: Walter de Gruyter GmbH, 43–58.

Winter, C. (2009a) 'Tourism, social memory and the Great War', *Annals of Tourism Research*, 36(4): 607–626.

Winter, C. (2009b) 'The Shrine of remembrance Melbourne: A short study of visitors' experiences', *International Journal of Tourism Research*, 11(6): 553–565.

Winter, C. (2010) 'Battlefield visitor motivations: Explorations in the Great War town of Ieper, Belgium', *International Journal of Tourism Research*, 13: 164–176.

Winter, J. (1985) *The Great War and the British People*, Basingstoke: Macmillan.

Winter, J. (1992) *Sites of Memory, Sites of Mourning: The Great War in European Cultural History*, Cambridge: Cambridge University Press.

Winter, J. (2006) *Remembering War: The Great War and Historical Memory in the Twentieth Century*, New Haven, CT: Yale University Press.

13 Reflecting on the *Great* War 1914–2019

How has it been defined, how has it been commemorated, how should it be remembered?

Brian S. Osborne

Introduction: constructing memories of the *Great* War

Remembering and commemorating, both informally and formally, have long been recognized as agencies in constructing an 'imagined community' and furthering identity and unity at the local, regional, and national levels (Anderson 1991). Through various devices, 'individuals are encouraged to recognize one another as members of a larger group, sharing a unified political space, and identifying with a common historical narrative of historical development' (Osborne 2002: 9). War has been a powerful agency in that process. Volumes of chronicles, galleries of heroic images, pantheons of monuments, and calendars of ceremonial extravaganzas have all been marshalled in campaigns choreographing the construction of national identities (Winter and Sivan 1999; Todman 2007; Doss 2010; Dolski 2016). National histories have been syncopated by the mythic accounts of iconic battles such as Agincourt, Gettysburg, Rorke's Drift, Stalingrad, and Waterloo, as have the mythic heroes associated with them. What would British history be without Shakespeare's *Henry V* and Agincourt, Wellington's 'nearest-run thing' of Waterloo, or 'Montgomery of Alamein'?

In this vein, the memory of that horrendous conflict, the *Great* War, is increasingly being reconstructed in the context of what has been called a 'memory boom' in which diverse collective meditations on war have been advanced by 'ubiquitous cultural projects', 'signifying practices', 'commemorative projects', and 'theatres of memory' that, when taken together, demonstrate the 'distance between history and memory' (Winter 2006: 1–2). If 'war made the state and the state made war' (Tilly 1975: 42), wars have played a fundamental role in defining and periodizing history and confusing national memory with patriotism (Kammen 1991: 13). More particularly, rather than being perceived as a tragedy to be avoided, war has often been advanced as a 'central force in forging nations and providing examples of valour for future citizens to remember and emulate' (Morgan 2016: 75; see also Moore and Whelan 2007).

Rather than being part of an historical chronicle of actual events, therefore, the history of warfare and military conflict has been manipulated by romantic mythologies and iconic symbols to encourage identification with the emerging concepts of nation and patriotism. Indeed, a recent study of the construction of national identity in the United States proposed that 'glorifying war-related deaths for political purposes' has a long pedigree in history and goes on to argue that 'historical remembrance affords a stage for multiple constituencies to stake claims on the past with implications for the present and the future' (Dolski 2016: 214). In other words, rather than being an academic exercise reflecting on the past, the study of warfare prompts a focus on its potential role in determining our futures.

This is the context which requires that our engagement with the *Great* War effects a critical evaluation of how it has been represented and interpreted. Indeed, there has been a complex array of motives and now, 100 years later, there is clearly a tension between the questionable sublimity of conflict and the putative moral objective of peace. Its centenary has prompted reflection on how we are *re*-membering the significance of that conflict and what we should be *co*-memorating for lessons learned. Hence, the focus on deconstructing the meanings of the *Great* War, der *Grosse* Krieg, le *Grand* Guerre, and the *Real* War (Liddell Hart 1963). The following questions are being advanced here as central to the present discussion: Why has the First World War been referred to as the *Great* War? How should we review it in the twenty-first century? What should we remember and what should we forget? In particular, should it be marked by the designation of UNESCO World Heritage Sites? If so, what are the Outstanding Universal Values ascribed to the *Great* War and how should they be communicated?

Defining the bigger picture

My main point is that the adjective *Great* is central to any understanding of the significance, meaning, and impact of the First World War. The term, *World* War, was first used in September 1914 by the German philosopher Ernst Haeckel, who claimed that 'there is no doubt that the course and character of the feared "European War" . . . will become the first world war in the full sense of the word' (www.famousscientists.org/ernst-haekel: accessed 1 July 2016). A month later, Canada's *Maclean's Magazine* declared, 'Some wars name themselves. This is the Great War'. Certainly, the eventual scale of the participation justified the descriptors of *Great* and *World*: over 50 nations participated; some nine million combatants and seven million civilians lost their lives; and the mass destruction of the rural and urban landscapes and their communities was immense (Winter and Baggett 1996). As the principal rationale for his study of the 'catastrophe' of World War I, Hastings speaks of it as commencing the 'grim march' of twentieth-century history marked as it was by 'sweeping military offensives', 'great pitched battles', 'staggering body-counts', and 'the stagnation of trench warfare' (Hastings 2013; see also Kershaw 2015). And the statistics of human loss *are* staggering! For Great Britain and her Dominions alone there were a million dead, 2.5 million wounded, 40,000 amputees, 60,000 blinded, 65,000 suffering from shell shock, and some

2.4 million were still disabled a decade later. In France, 75 per cent of the male population between 18 and 30 years of age were killed or wounded (Lane, cited in Betts 2015: 156–157).

Scale aside, the conflict is also associated with such major socio-political developments that it prompted the proposition that 1914 marked the end of the 'long' nineteenth century (Hobsbawm 1987). What emerged out of the trenches was a growing 'modernity' accompanied by a nihilism and alienation that would 'eventually crush social structures, moral norms, and traditional ideas' (Blom 2015: 406; see also Eksteins 1989). But if the *Great* War was a profound demonstration of social and intellectual disruptions associated with modernity, there were also geopolitical implications that were immense and portentous for the following century. As MacMillan put it in her analysis of the 'peace' following the Treaty of Versailles, with the collapse of tsarist Russia, the Austria-Hungarian and Ottoman empires, '[o]ld nations – Poland, Lithuania, Estonia, Latvia – came out of history to live again, and new nations – Yugoslavia and Czechoslovakia – struggled to be born' (2001: xxvi). The collapse of many transnational empires and the emergence of 'nation-states' defined by language or declared ethnicity was a new verity realized by the assertion of 'national self-determination' at Versailles.

Indeed, the geopolitical-myopia that accompanied the cartographic exercises identifying ethnic-linguistic boundaries in the Hall of Maps at Versailles 'has cast a long shadow over the twentieth century, and beyond' so that 'the horrors of the conflict still commands attention, and will probably never cease to do so' (Otte 2014: 1). And, a century later, the world is reacting to ongoing repercussions and ramifications of past misconceived macro-geopolitics and current micro-identities. If the 'short' twentieth century strove to achieve international cooperation, transnational linkages, and the dream and rhetoric of cosmopolitan unity, the twenty-first century is still immersed in struggles over national identities, frontiers, and homelands in Europe, Asia, the Americas, and the Middle East that are so reminiscent of the prelude to the *Great* War.

Constructing the 'place' of the *Great* War

Central to this label of 'greatness' is the affect warfare had on particular 'places'. As geographers, we have long become sensitive to the power of this relationship and its expression in the 'landscape' around us and how it has been created. In considering how this has come to be, we see landscape as a complex surface consisting of successive layers of 'sediments of meaning' which remain as powerful 'latent presences' (Walsham 2011: 564). That is, rather than being simply material representations of past economic, social, and cultural processes, landscapes serve as documents that record layers of meaning in this geographical palimpsest that are communicated in diverse ways (Sørensen and Viejo-Rose 2015). Recognition of this process has prompted several neologisms that focus on dominant processes involved in the construction of the essence of places: 'warscapes', 'inscapes', 'memoryscapes', 'prosthetic-scapes', and one central to this study, 'proscapes'.

Warscapes

The most enduring and commonly accepted iconic image of the place of the *Great War* is that of the actual physical expression of the locus of conflict, the 'warscape' experienced by millions of combatants (Todman 2007). While the war was fought in the air, in naval battles worldwide, and in martial conflicts in the Middle East, Africa, and Asia, it is the Western Front that bulks large in the popular memory of the *Great* War. Over centuries, quotidian economic, societal, and cultural activities had produced a material landscape that reflected the lived values and priorities of communities in these places. But, in a narrow corridor of warfare stretching from the Baltic to the Alps, these urban and rural worlds were destroyed by the intensity of modern warfare. Traditional *paysages* and *landschafts* of farmsteads, villages, and towns were replaced by a dystopian 'warscape' of trenches, craters, and barbed wire in a lived-in world of explosions, machine-gun fire, shellfire, and clouds of gas. Perhaps an important explanation here is a combination of the locational fixity of the trench warfare, and the immense destruction effected by the power of industrialized war. The enormity of the colossal transformation of place is demonstrated by the prediction advanced by Major General Charles Harington on the eve of the detonation of the several massive subterranean mines that marked the opening of the Battle of Messines in June 1917: 'Gentlemen, we may not make history tomorrow, but we shall certainly change the geography' (Passingham 1998: 90)! The accuracy of Harington's prescience is demonstrated by the cratered landscape that remains to this day as a landmark of the *Great* War.

The effect of war was also felt beyond the trenches and fire-lines. Apart from being a regrettable consequence of the conduct of war, the deliberate targeting of a society's totemic monuments and domestic space was often a strategy to erode cultural memory and, therefore, morale (Bevan 2016). A recent biography of Edith Wharton captures her reaction to the destruction of Ypres as written in June 1915:

> We had seen other ruined towns, but none like this. The towns of Lorraine were blown up, burned down, deliberately erased from the earth. At worst they are like stone-yards, at best like Pompeii. But Ypres has been bombarded to death.
>
> (Kimber 2016: 10)

It has been estimated that in the 500 km long 'zone of death' of the Western Front some 3.3 million hectares where the population had decreased by 44 per cent, and some 800,000 buildings and 1,954 settlements had been destroyed or badly damaged (Clout 1996: 19–58).

However, while much of the current retrospection of the war is dominated by these images of combat, devastation, and male-sacrifice, others have argued that a fuller appreciation of the *Great* War necessitates 'climbing out from the perspective of the trenches' and address the impact of the war on gender, class, service, and new values on the home front (Watson 2004: 10–13). Beyond the verities of combat on the battlefields, there were also dramatic changes in life at home: employment opportunities for, and enfranchisement, of women; the disenfranchisement of recent immigrants from Germany and Austro-Hungary; as volunteering proved insufficient, the imposition of conscription; the organization of labour; and pacifism. These too should be recognized as 'warscapes' that capture the changes produced by the *Great* War.

Inscapes

Such was the mass-horror and disruption of the *Great* War that the experience of it was rendered through art, literature, theatre, and film in reflexive images of the conflict. Not only did these record the physical conversion of pre-war normalcy into a dystopic place, they also generated profound sensory encounters with a world transformed. The hitherto neutral names of Ypres, Verdun, Somme, Passchendaele, and Gallipoli became associated with the essence of a conflict captured and interpreted through emotional, ethical, moral, and didactic lenses. These constructed narratives and images of the sites/sights of the war continue to create inscapes that evoke the conduct, experience, and lessons learned (Tippett 1984; Mosse 1990; Winter 1995, 2006; Osborne 2001b, 2001c; Brandon 2006, 2007).

But the visual representations of the *Great* War evolved as the conflict persisted and evolved from the style of the heroic art of British colonial wars to later depictions of the destruction and horror of the trenches, the grievously wounded, and the dead (Oliver and Brandon 2000: xiii). However, some war artists initially saw the *Great* War as 'a thrilling quest for human advancement' until the reality of twentieth-century warfare soon revealed a different picture, 'one so ghastly that no one could completely capture it', although some struggled to do so and 'their world would never look at itself in the same way again' (Castel 2004: 9).

Apart from grappling with the artistic challenge posed by the new warfare, others like C.R.W. Nevinson proposed that as 'there was no beauty except in strife, and no masterpiece without aggressiveness' that war would have a powerful effect on the cultural agenda of nations (Fox 2015: 4). This was certainly the case for Canada. The grim realities of conflict increasingly attracted the attention of battlefield artists and gradually, the pre-war artistic national self-image that centred on impressionistic representations of a bucolic rural landscape was replaced by renderings of a people-less and challenging north that demanded human sacrifice (Tippett 1984; Osborne 1988; Lord 1974). Elsewhere, others were influenced by the reality of, and reactions to, modern warfare. They rendered both the new technology of trenches, barbed wire, and gas while also posing ironic questions and cynical answers for the intent of the conflict. And across other side of no-mans-land, Otto Dix had the same agenda with his horrific interpretations of combat and post-war suffering of German veterans (Hermann 2010).

Nor was the dislocation and suffering of the displaced civilian population neglected. Artists were to the fore in recognizing the social consequences of the war on its place and people as displaced families faced the grim reality of post-war recovery and reconstruction of their devastated homes and communities. Of all the artists who have contributed to the visual inscapes of this process, it was Mary Riter Hamilton, a woman artist in a field dominated by men, who captured the material and psychological damage immediately after the end of hostilities (Davis and McKinnon 1992; Osborne 2001b, 2001c). From 1919 to 1922, Hamilton produced over 300 images of the abandoned battlefields, devastated communities, and the return of the civilian population to their former homes and lives. Her rendering of people-less, transformed, and silent landscapes such as *The Kemmel Road, Flanders* (1920) and *Trenches on the Somme* (1919) provoke reflection on past suffering while her *First Boat to arrive in Arras after the Armistice* (1920) and *Market among the ruins of Ypres* (1920) communicate the promise of recovery as a dominant blasted cityscape is being recolonized by clusters of returning people.

In another imaginative domain, literary figures such as Rupert Brooke, Robert Graves, Wilfred Owen, Erich Maria Remarque, Siegfried Sassoon, and many others, were prominent symbolic interpreters of the conflict. Their communication of the realities and emotional impact of modern warfare in books such as Graves' *Goodbye to All That* and Remarque's *All Quiet on the Western Front* became diagnostic of a counter-narrative to wartime patriotism. Perhaps the most widely recognized patriotic panegyric of the *Great* War was John McCrae's *In Flanders Fields* which, while appreciated by many for its powerful poetic imagery in 1915, has also been critiqued by others in later times for its appropriation by others for political purposes and even commodification (Betts 2015). McCrae's evocative reference to the place of war is still the imagery of '*In Flanders fields the poppies blow; Between the crosses, row on row*'. But his invocation to future generations to '*Take up our quarrel with the foe*' has long been used by many for militaristic promptings. Others do urge understanding McCrae's verses in the context of the time of their creation and not subject his imagery and message to 'presentist' interpretations. But some poets *did* react differently to the verities of war at the time. Wilfred Owen's *Dulce et Decorum est* was penned sometime between October 1917 and March 1918 and, while not as popular at the time, it captured what so many actually experienced at the front. Such imagery as '*blood gargling from the froth-corrupted lungs*' and '*vile, incurable sores on innocent tongues*' prompted him to urge others to '*not tell with such high zest to children ardent for some desperate glory*' that it is '*Dulce et Decorum est Pro patria mori*'.

Taken together, these 'inscapes' have left indelible images of the emotional reaction to the *Great* War and, in doing so, contributed so much to policies and practices of remembering its consequences.

Memoryscapes

While some artists of the *Great* War produced actual 'warscapes' that were subsequently rendered by others in impressionistic 'inscapes', the actual place and conduct of the *Great* War has also been communicated over the last century by constructed 'memoryscapes' that constitute ritualized reminders of notable battles and the general suffering, sacrifice, and mourning associated with the *Great* War.

Often located at actual sites of battles and of heroic loss and achievement, they syncopate remembrance at dedicated sites of commemoration according to a calendar of daily, monthly, and annual ceremonies.

Monument-centred remembrance of war has transformed much of the former warscape of the Western Front into a memorialised landscape where '[w]ar memorials dot the countryside, in cities, towns, and villages, in market squares, churchyards, schools, and obscure corners of hillsides and fields' (Winter 1995: 1).These commemorative constructions have the objective of honouring the fallen and their sacrifice, rendering it as noble and heroic, and thus assuaging the loss for those left behind. Ideally, they were constructed to 'instill a sense of historical closure' and 'commemoration was a process of condensing the moral lessons of history and fixing them in place for all time' (Savage 1999: 14).

But, like other interventions in the historical record, monuments and commemoration have evolved over time, as have the messages they were meant to communicate. Often structurally impressive visual presences, these monuments are intended to be viewed by individuals and groups. As such, they are intended to contribute to collective memories by promoting a shared experience of the values and emotions communicated in their material form – often loaded with the triumphalist jingoism of the victors tied to particular nationalist projects (Moriarty 1995, 1999; King 1998, 1999a; Ashplant et al. 2000; Osborne 2001a; Johnson 2003; Nelson and Olin 2003; Shanken 2004; Winter 2009; McKay and Swift 2012, 2016). Such a perspective integrates the intrinsic power and 'concretized memory' of monuments into 'a nation's rites', become 'objects of a people's national pilgrimage' and are often 'stubbornly resistant to the state's original intentions' (Young 1993: 2–3). This critical assessment of the original purpose and actual role of monuments in memory-construction must, therefore, always be to the fore. Attention must also be directed to the danger of nurturing an imagined historical space of 'mythscapes' which are invented narratives created and transmitted for posterity (Moriarty 1995; Bell 2003).

While outnumbered by the hundreds of memorials commemorating the wartime achievements of nations, regiments, and individual heroes, other initiatives of memorialisation have been specifically dedicated to the end of war and the prospect of post-war peace and reconciliation. Often, those who had served and returned wanted to mark the service and sacrifice of those who hadn't by making a contribution to their home community in their memory. While they too often took the form of monumental towers, arches, or statues, others chose more pragmatic contributions such as memorial hospitals, universities, community halls, gardens, bridges, and highways (Lollis 2013).

Some of these innovative memory-projects have looked back to the actual 'place' of conflict. The very nature of the *Great* War had destroyed the *genre de vie*, the quotidian landscape, and symbolic architecture of the pre-war society and some memorial initiatives focused on the post-war reconstruction of the lived-in worlds destroyed by war (Bevan 2016). By 1922, the population of France's war zone had returned to 88.6 per cent of pre-war levels, but most occupied temporary housing as accommodation and services had yet to be restored (Osborne 2001c:

65–66). The objective was clear: the 'restored countryside' was to become a statement of the reassertion of civil order and normality by the 'replacement of warscapes by the vernacular artefacts of domesticity and community – and the obliteration of the memory of war' (Osborne 2001c: 66). Another remarkable dimension was the relationship that had been developed by British troops with the devastated communities. If the stasis and fixity of the conflict had generated established trench-lines, it had also nurtured contact with the local communities. The result had been the forging of strong emotional links between returned military units and the French communities with which they had interacted. This was the context of the 'British League for the Reconstruction of the Devastated Areas of France', which was founded on 30 June 1920 (Clout 1996; Osborne 2001a, 2001b, 2001c). By July 1922, 78 British and colonial communities had adopted 99 communities and contributed to the restoration of homes, churches, civic buildings, and such pragmatic projects as water supplies. The plaques celebrating these projects and their motivation are also elements of the memoryscapes of the *Great War*, several of which later served as linkages perpetuated as 'paired communities' in the European Union.

Prosthetic-scapes

A prominent initiative in the remembrance and co-memoration of the *Great* War over the last century has been what Landsberg (2004) calls 'prosthetic-scapes': the generation of museums, galleries, and guided tours to reconstructed battlefields. Immediately following the cessation of hostilities, battlefield tourism commenced with what amounted to virtual pilgrimages. The initial visits were organized as patriotic visitations by veterans, widows, and orphans, and some as didactic experiences for schoolchildren (Iles 2008; Lloyd 1998). Over time, while these original motivations continued, they were joined by others seeking an enhanced experiential encounter ranging from the 'dark experiences' of thanatourism: the focus on sites of tragedies, disasters, and death by those wishing to learn more from enhanced and interpretive representations of places of atrocity, disasters, and suffering (Seaton 2000; Dunkley et al 2011; Hertzog 2012; Hartmann 2014; Miles 2014). The battlefields of the *Great* War certainly fit this category and attract many.

Accommodating these new demands has generated a 'prosthetic-scape' by the presentation of the verities of the *Great* War through different forms of media. Battlefields have long been represented by messages communicated by glass-cabinets of artefacts, panels of photographs and texts, and the voiced commentary of traditional museums. Increasingly, these are now accompanied, or even replaced, by cinematic or virtual-reality simulations (Landsberg 2004, 2015). Such new technologies add an innovative and powerful dimension to the manipulation of memory and heritage. Building on the observation of Senator Hiram Johnson Senator to the United States Senate in 1917, that the 'first casualty when war comes is truth', Andrew Hunter opens his study, *Dark Matter: The Great War and Fading Memory*, with a provocative allusion to how our memories are presently being constructed and, perhaps, misrepresented; '[w]e live in a time of exaggerated feelings, where historical events are used . . . repackaged and remessaged to frame the present' (Hunter 2007: i). His point is that, more than ever, the memory of that horrendous conflict 'seems detached from whatever truth remains, if it ever existed in the first place'.

While tourism attracted by such representation has been accused of the 'desacrilization' of commemoration, it does offer a powerful dimension in the relationship between identity and memory (Carsten 2007). Indeed, recent innovations in 'commemorative practice' have suggested new opportunities in the centenary celebrations of the *Great* War. On the one hand, the new concern with the 'experiential economy' is directing attention to the potential of enhanced tourism for economic growth (Kovacs and Osborne 2008). The economic benefits of tourism have long been associated with culture, history, and heritage, whilst increasing investment in the 'battlefield tourist industry' now makes a significant contribution to the 'regional and national economy' (Vanneste and Foote 2013: 255). While Ypres has adopted the title and role of 'The City of Peace', attention should be directed to the propriety of such initiatives as the 'British Grenadier Bookshop', 'The Old Bill Pub', and 'The Poppy Pizzeria and Steakhouse'. The evolving cultural environment of the *Great* War is increasingly dominated by this re-packaging and re-messaging as a new form of public cultural memory that is being introduced into identity politics. And some have an optimistic expectation of the legacy of the *Great* War and that its centenary 'might become a symbol of respectful remembrance and of building global values in a new world' and that tourism becomes 'a driving force for sharing values, for education and knowledge, for international solidarity, and for passing on the torch of peace' (Jansen-Verbeke and George 2013: 285). That is, as Vanneste and Foote (2013: 256) argue, there is an opportunity to move 'prosthetic-scapes' beyond regional economics and to re-orientate 'battlefield (and dark) tourism to peace (and heritage) tourism'.

Constructing Proscapes

Given the striking impressions of the realities of 'warscapes', the emotive impressions of 'inscapes', the reinforcement of the experience in 'memoryscapes' and 'timescapes', and the mass exposure to all of these through 'prosthetic-scapes', it is not surprising if, a century later, some are questioning what we have learned. In particular, it is being argued that we should be establishing 'proscapes' that embrace new ethics, priorities, and values, and promote them by new interventions. That is, energizing the forces of memory into a view of the future that embraces new hopes that will serve as an 'ethics of remembering', and 'intangible' milestones that serve as signposts for progress to a better future (Wyschogrod 1998; Boime 1998).

Nor is this a new concern. Immediately following the termination of the *Great* War, there was considerable controversy over what should be commemorated and what should be learned. For some, the emphasis was on triumphalism and military recognition; for all, it was bereavement, and for a few, it was on lessons to be remembered. In his analysis of the meaning of post-war monuments, King quotes war correspondent, Phillip Gibbs, who declared soon after the *Great* War that the real purpose of war monuments was 'the safeguard of the living by teaching those who follow to learn wisdom by our stupidity, and to cherish the gift of peace' (cited in King 1999b: 7). King agrees and, while recognising the long-standing debate on the effectiveness of memorials, also asserts that their purpose should be as 'reminders of the horror, waste and ultimate futility of war, and as important devices in the campaign to prevent war occurring again' (King 1999b: 7). Clearly, how and what we remember has always been a contested terrain in which the complexity of personal, community, and national preferences ensure diverse outcomes.

Thus, while many of the *Great* War memorials and commemorations were aimed at celebrating the sacrifice of the victors, there have been other initiatives to promote a better world by learning from the sacrifice, suffering, and loss experienced by all participants, combatants and civilians. Perhaps one of the earliest of these gestures of reconciliation between former foes was prompted by one of the bloodiest campaigns in the *Great* War. In 1934, Kemal Ataturk, the first president of modern Turkey, penned a tribute to all those who died at Gallipoli:

> Those heroes that shed their blood and lost their lives . . . You are now lying in the soil of a friendly country. Therefore rest in peace. There is no difference between the Johnnies and the Mehmets to us where they lie side by side here in this country of ours . . . You, the mothers who sent their sons from faraway countries, wipe away your tears; your sons are now lying in our bosom and are in peace. After having lost their lives on this land they have become our sons as well'.
>
> (www.awm.gov.au/encyclopedia/ataturk, accessed 16 June 2016)

However, elsewhere, the concept of conciliation could not overcome the momentum of wartime prejudice and propaganda. One of the major obstacles to post-war reconciliation was the post-war assignment of responsibility for initiating the conflict and exacting retribution was the 'war guilt clause' confirmed in Article 231 of the Treaty of Versailles (MacMillan 2001: 193). But recently, there have been initiatives to overcome this. In Germany, the remembrance event of *Volkstrauertag* grew out of the tradition of commemorating all victims of war rather than simply the German war dead. In Kitchener, Ontario, this annual ceremony of 'Shared Grief' has become the 'People's Day of Mourning' (Kaiser 2010; Kovacs 2016). And it is growing. In 2014, the ceremony marking commencement of the *Great* War a century earlier featured an 'All Friends at Last' ceremony of reconciliation on the battlefield of the Western Front (Nicholas 2013). Increasingly, in Australia, New Zealand, and elsewhere, Anzac Day is incorporating the recognition of the suffering and losses on both sides in the horrific conflict at Gallipoli in that 'All graves are one' (Inglis 1998: 483).

Conclusion: the quest for universal values

In conclusion, it is clear that any consideration of the *Great* War a century later must attempt to evaluate it in the context of the time frame of that period. But, while recognizing the danger of a 'presentist' agenda, we must also strive to seek insights today for the lessons learned from the past and current tensions.

So, is it possible to conceive of a reflexive transnational 'proscape' located in the battlefields and beyond of the *Great* War? Much of the aforementioned discussion has addressed mostly national, and some bi-national, initiatives such as the British League for Reconstructing the Devastated Areas of France, Germany's *Volkstrauertag*, and Atatürk's promotion of peace and understanding between nations. But the very nature of present and future international tensions requires a truly global perspective. It is in this context that UNESCO recognizes that certain locales should be inscribed on the World Heritage List for their 'Universal Heritage Values' as defined in Paragraph 49 of the Operational Guidelines:

> Outstanding universal value means cultural and/or natural significance which is so exceptional as to transcend national boundaries and to be of common importance for present and future generations of all humanity. As such, the permanent protection of this heritage is of the highest importance to the international community as a whole.
> (http://whc.unesco.org/archive/opguide05-en.pdf: accessed 16 June 2016)

Accordingly, World Heritage Sites (WHS) have been established to serve as signposts for human progress in culture, politics, and social organization (Di Giovine 2008). In 1981, UNESCO revisited Atatürk's promotion of peace and understanding between the nations of the world and adopted the *Resolution on the Atatürk Centennial* (*www.turkishnews.com/Ataturk/unesco.htm:* accessed 4 July 2016.)

But Christina Cameron thinks we should go further. She recognizes that the UNESCO World Heritage Convention 'deals more comfortably with the physical manifestations of heritage places' and that 'sites of conscience present a challenge to determine the threshold of outstanding universal value for their intangible dimensions' (Cameron 2010: 113). Sites such as Auschwitz-Birkenau and the Hiroshima Peace Memorial serve as 'poignant reminders of past injustices and powerful catalysts for making the world a better place' (Cameron 2010: 117). In conclusion, she argues that these 'sites of conscience' have 'powerful intangible values that recall human rights abuses and other injustices' and, by contributing to 'sustaining memories and understanding the roots of conflict' they can foster a broad-based contemporary dialogue that 'underscores their global importance' (Cameron 2010: 118).

What would those new tropes of memory be? While recognizing that past history and present tensions demonstrate that there have been 'good wars' (Dolski 2016: 15), there are lessons to be learned from how we have approached the *Great* War. Reflecting on the multiple landscapes of meaning generated by the conflict, several approaches to its remembrance and commemoration have been suggested here:

* *recognition* of its unique scale, technology, practice, and consequences;
* *rejection* of militaristic and patriotic manipulations of state conflict;
* *appreciation* of the human sacrifice and suffering caused by war;
* *celebration* of the significant literary, artistic, and philosophical reflections on war; and
* *promotion* of the abhorrence of war and the advocacy of international cooperation as an ideal.

Ideally, these objectives should promote a view of the future that is emancipated from institutionalized nationalist prejudices and aims at nurturing a transnational mind-set that celebrates the values of a cosmopolitan world. In this vein, Margaret MacMillan's concluding reflections on the aftermath of the putative peace constructed at Versailles and the end of the *Great* War are provocative: 'How can the irrational passions of nationalism or religion be contained before they do more damage? How can we outlaw war' (MacMillan 2001: 494)? Answering these 'big questions' would truly make the First World War *Great*!

Bibliography

Anderson, B. (1991) [1983] *Imagined Communities: Reflections on the Origins of Nationalism*, London: Verso.

Ashplant T.G., Dawson, G. and Roper, M. (eds.) (2000) *The Politics of War Memory and Commemoration*, London: Routledge.

Bell, D. (2003) 'Mythscapes: Memory, mythology and national identity', *British Journal of Sociology*, 54(1): 63–81.

Betts, A. (ed.) (2015) *In Flanders Fields: 100 Years*, Toronto: Knopf Canada.

Bevan, R. (2016) *The Destruction of Memory: Architecture at War*, London: Reaktion Books.

Blom, P. (2015) *Fracture: Life and Culture in the West, 1918–1938*, New York: Basic Books.

Boime, A. (1998) *The Unveiling of the National Icons: A Plea for Patriotic Iconoclasm in a Nationalist Era*, Cambridge: Cambridge University Press.

Brandon, L. (2006) *Art or Memorial? The Forgotten History of Canada's War Art*, Calgary: University of Calgary Press.

Brandon, L. (2007) *Art and War*, New York: I.B. Tauris.

Butler, R. and Suntikul, W. (2013) *Tourism and War*, London: Routledge.

Cameron, C. (2009) 'The evolution of the concept of Outstanding Universal Value', in N. Stanley-Price and J. King (eds.), *Conserving the Authentic: Essays in Honour of Jukka Jokilehto*, Rome: ICCROM Conservation Studies, 127–136.

Cameron, C. (2010) 'World heritage sites of conscience and memory', in D. Offenhäuser, W. Zimmerli and M.-T. Albert (eds.), *World Heritage and Cultural Diversity*, Cottbus: German UNESCO Commission, 112–119.

Carsten, J. (ed.) (2007) *Ghosts of Memory: Essays on Remembrance and Relatedness*, Malden, MA: Blackwell.

Castel, B (2004) 'From the Editor: The Spirit of Our Time', *Queens Quarterly*, 111(1): 9.

Clout, H. (1996) *After the Ruins: Restoring the Countryside of Northern France after the Great War*, Exeter, UK: Exeter University Press.

Davis, A. and McKinnon, S. (1992) *No Man's Land: The Battlefield Paintings of Mary Riter Hamilton, 1919–1922*, Winnipeg: University of Winnipeg.

Di Giovine, M.A. (2008) *The Heritage-Scape: UNESCO, World Heritage, and Tourism*, Lanham: Lexington Books.

Dolski, M.R. (2016) *D-Day Remembered: The Normandy Landings in American Collective Memory*, Knoxville: University of Tennessee Press.

Doss, E. (2010) *Memorial Mania: Public Feeling in America*, Chicago: University of Chicago Press.

Dunkley, R., Morgan, N. and Westwood, S. (2011) 'Visiting the trenches: Exploring meanings and motivations in battlefield tourism', *Tourism Management*, 32(4): 860–868.

Eksteins, M. (1989) *Rites of Spring: The Great War and the Birth of the Modern Age*, Toronto: Lester & Orpen Dennys.

Fox, J. (2015) *British Art and the First World War, 1914–1924*, Cambridge: Cambridge University Press.

Foote, K.E. and Vanneste, D. (2011) 'The Menin Gate Memorial, Ieper', in P. Post, A.L. Molendijk and J.E.A. Kroesen (eds.), *Sacred Places in Modern Western Culture*, Leuven, BE: Peeters Publishers, 253–257.

Hartmann, R. (2014) 'Dark tourism, thanatourism, and dissonance in heritage tourism management: New directions in contemporary tourism research', *Journal of Heritage Tourism*, 9(2): 166–182.

Hastings, M. (2013) *Catastrophe 1914: Europe Goes to War*, New York: Alfred A. Knopf.

Hermann, I. (2010) *Biography: Otto Dix*, edited by O. Peters, New York: Prestel Publishing.

Hertzog A. (2012) 'War, battlefields, tourism and imagination', *International Interdisciplinary Review of Tourism*, 1: 1–13.

Hobsbawm, E. (1987) *The Age of Empire: 1875 – 1914*, UK: Weidenfield & Nicolson.

Hunter, A.T. (2007) *Dark Matter: The Great War and Fading Memory*, Smithville, ON: Carruthers Printing.

Iles J. (2008) 'Encounters in the fields: Tourism to the battle fields of the Western Front', *Journal of Tourism and Cultural Change*, 6(2): 138–154.

Inglis, K.S. (1998) *Sacred Places: War Memorials in the Australian Landscape*, Melbourne: Melbourne University Press.

Jansen-Verbeke, M. and George, W. (2013) 'Reflections on the Great War centenary: From warscapes to memoryscapes in 100 years', in R. Butler and W. Suntikul (eds.), *Tourism and War*, London: Routledge, 273–287.

Johnson, N (2003) *Ireland, the Great War and the Geography of Remembrance*, Cambridge: Cambridge University Press.

Kaiser, A. (2010) 'The *Volkstrauertag* (People's Day of Mourning) from 1922 to the present', in B. Niven and C. Paver (eds.), *Memorialization in Germany since 1945*, New York: Palgrave Macmillan, 15–25.

Kammen, M. (1991) *Mystic Chords of Memory: The Transformation of Tradition in American Culture*, New York: Vintage Books.

Kershaw, I. (2015) *To Hell and Back: Europe, 1914–1949*, New York: Viking.

Kimber, G. (2016) 'Review of Alice Kelly (ed.) "Fighting France: From Dunkerque to Belfort"', *The Times Literary Supplement*, 10 June, p. 10.

King, A. (1998) *Memorials of the Great War in Britain: The Symbolism and Politics of Remembrance*, Oxford and New York: Basic.

King, A. (1999a) 'Remembering and forgetting in the public memorials of the Great War', in A. Forty and S. Küchler (eds.), *The Art of Forgetting*, Oxford: Berg, 147–169.

King, A. (1999b) 'Monuments with no fixed meaning', *The Independent*, 23 March, p. 7.

Kovacs, J. (2016) 'War remembrance in a sacralized space of memory: The origins and evolution of *Volkstrauertag* in Kitchener, Ontario, Canada', *Memory Studies*, 9(2): 218–234.

Kovacs, J. and Osborne, B.S. (2008) 'Cultural tourism: Authenticity, escaping into fantasy, or experiencing reality: Bibliographic essay', *Choice*, 45(6): 927–939.

Landsberg, A. (2004) *Prosthetic Memory: The Transformation of American Remembrance in the Age of Mass Culture*, New York: Columbia University Press.

Landsberg, A. (2015) *Engaging the Past: Mass Culture and the Production of Historical Knowledge*, New York: Columbia University Press.

Liddell Hart, B.H. (1963) [1930] *The Real War 1914–1918*, New York: Little, Brown.

Lloyd, D. (1998) Battlefield Tourism: Pilgrimage and the Commemoration of the Great War in Britain, Australia and Canada, New York: Berg.

Lollis, E. (2013) *Monumental Beauty: Peace Monuments and Museums around the World*, Knoxville, TN: Peace Partners International.

Lord, B. (1974) *The History of Painting in Canada: Toward a People's Art*, Toronto: New Canada Publications.

MacMillan, M. (2001) *Paris 1919: Six Months That Changed the World*, New York: Random House.

MacMillan, M. (2013) *The War That Ended Peace: The Road to 1914*, Toronto: Allen Lane.

McKay, I. and Swift, J. (2012) *Warrior Nation: Rebranding Canada in an Age of Anxiety*, Toronto: Between The Lines.

McKay, I. and Swift, J. (2016) *The Vimy Trap: Or, How We Earned to Stop Worrying and Love the Great War*, Toronto: Between The Lines.

Miles, S. (2014) 'Battlefield sites as dark tourism attractions: An analysis of experience', *Journal of Heritage Tourism*, 9(2): 134–147.

Moore, N. and Whelan, Y. (eds.) (2007) *Heritage, Memory and the Politics of Identity: New Perspectives on the Cultural Landscape*, Farnham, UK: Ashgate.

Morgan, C. (2016) *Commemorating Canada: History, Heritage, and Memory*, Toronto: University of Toronto Press.

Moriarty, C. (1995) 'The absent dead and figurative First World War memorials', *Transactions of Ancient Monuments Society*, 39: 7–40.

Moriarty, C. (1999) 'The material culture of Great War remembrance', *Journal of Contemporary History*, 34(4): 633–662.

Mosse, G (1990) *Fallen Soldiers: Reshaping the Memory of the World Wars*, New York: Oxford University Press.

Nelson, R. and Olin, M. (eds.) (2003) *Monuments and Memory, Made and Unmade*, Chicago: Chicago University Press.

Nicholas, H. (2013) 'All friends at last on the western front: Britain and Germany will mark the start of the Great War with an act of reconciliation on the battlefield', *The Sunday Times*, 28 April, p. 13.

Oliver, D.F. and Brandon, L. (2000) *Canvas of War: Painting the Canadian Experience, 1914–1945*, Vancouver: Douglas & McIntyre.

Osborne, B.S. (1988) 'The iconography of nationhood in Canadian Art', in D. Cosgrove and S. Daniels (eds.), *Iconography of Landscape*, Cambridge: Cambridge University Press, 162-178.

Osborne, B.S. (2001a) 'Warscapes, landscapes, inscapes: France, war, and Canadian national identity', in I. Black and R. Butlin (eds.), *Place, Culture, and Identity*, Quebec: Laval University Press, 311–333.

Osborne, B.S. (2001b) 'Erasing memories of war: Reconstructing France after the "Great War"', in Y. Tremblay (ed.), *Canadian Military History since the 17th Century/L'Histoire Militaire Canadienne depuis le XVIIe Siècle*, Ottawa: National Defence, 513-522.

Osborne, B.S. (2001c) 'In the shadows of monuments: The British League for the recon-struction of the devastated areas of France', *International Journal of Heritage Studies*, 7(1): 59–82.

Osborne, B.S. (2002) 'The place of memory and identity', *Diversities*, 1(1): 9–13.

Otte, T. (2014) *July Crisis: The World's Descent into War, Summer 1914*, Cambridge: Cambridge University Press.

Passingham, I. (1998) *Pillars of Fire: The Battle of Messines Ridge, June 1917*, Stroud: Sutton Publishing.

Post, P., Molendijk, A.L. and Kroesen, J.E.A. (eds.) (2011) *Sacred Places in Modern West-ern Culture*, Leuven: Peeters.

Savage, K. (1999) 'The past in the present: The life of memorials', *Harvard Design Maga-zine*, Fall: 14–19.

Seaton, A.V. (2000) 'Another weekend away looking for dead bodies: Battlefield tourism on the Somme and in Flanders', *Tourism Recreation Research*, 25(3): 63–77.

Shanken, A.M. (2004) 'Research on memorials and monuments', *Annales des Instituto de Investigaciones Estéticas*, 84: 163–172.

Sørensen, M. and Viejo-Rose, D. (eds.) (2015) *War and Cultural Heritage: Biographies of Place*, New York: Cambridge University Press.

Tilly, C (1975) *The Formation of National States in Western Europe*, Princeton: Princeton University Press.

Tippett, M. (1984) *Art at the Service of War: Canada, Art, and the Great War*, Toronto: University of Toronto Press.

Todman, D. (2007) *The Great War: Myth and Memory*, London: Bloomsbury Academic.

Vance, J.F. (1997) *Death So Noble: Memory, Meaning, and the First World War*, Vancou-ver: University of British Columbia Press.

Vanneste, D. and Foote, K. (2013) 'War, heritage, tourism and the centenary of the Great War in Belgium', in R. Butler and W. Suntikul (eds.), *War and Tourism*, Oxford and New York: Routledge & Taylor and Francis, 254–272.

Walsham, A. (2011) *The Reformation of the Landscape: Religion, Identity, and Memory in Early Modern Britain and Ireland*, Oxford: Oxford University Press.

Watson, J.S.K. (2004) *Fighting Different Wars: Experience, Memory, and the First World War in Britain*, Cambridge: Cambridge University Press.

Winter, C. (2009) 'Tourism and social memory and the Great War', *Annals of Tourism Research*, 36(4): 607–626.

Winter, C. (2011) 'Battlefield visitor motivations: Explorations in the Great War town of Ieper, Belgium', *International Journal of Tourism Research*, 13(2): 164–176.

Winter, C. (2012) 'Commemoration of the Great War on the Somme: Exploring personal connections', *Journal of Tourism and Cultural Change*, 10(3): 248–263.

Winter, J. (1995) *Sites of Memory, Sites of Mourning: The Great War in European Cultural History*, Cambridge: Cambridge University Press.

Winter, J. (2006) *Remembering War: The Great War between Memory and History in the Twentieth Century*, New Haven and London: Yale University Press.

Winter, J. and Baggett, B. (1996) *The Great War and the Shaping of the 20th Century*, London: Penguin.

Winter, J. and Sivan, E. (eds.) (1999) *War and Remembrance in the Twentieth Century*, Cambridge: Cambridge University Press.

Wyschogrod, E (1998) *An Ethics of Remembering: History, Heterology, and the nameless others*, Chicago: University of Chicago Press.

World Heritage Tourism Research Network (2012). Available from: www.whtrn.ca.

Young, J (1993) *Texture of Memory: Holocaust Memorials and Meaning*, New Haven and London: Yale University Press.

14 Afterword

The mobilization of memory
1917–2014

Paul Cornish

The papers in this volume leave the reader in no doubt as to the range of structures and landscapes which can be considered commemorative spaces of the First
World War. They can be as circumscribed as a plaque on a wall, or as extensive
as the whole of the Western Front. Indeed, even as the war was raging, the Front
itself was being reconfigured as a site of memory. While soldiers marked the
graves of their comrades and erected monuments to them, artists such as Paul
Nash and CRW Nevinson, or photographers like Frank Hurley, interpreted the
battlefields in a way that has formed the foundation of our impression of them to
this day (Gough, this volume, Osborne, this volume). Soon after the war's end
the erstwhile Front became the venue for 'pilgrimages' (Pennell this volume)
and (along with the recently christened Home Front) the host to a range of commemorative structures. Over the ensuing years, some strange juxtapositions have
resulted; for example, the stark whiteness of the Canadian memorial's Croatian
limestone, embedded in the loess of Vimy Ridge (Foster this volume). Likewise,
the incongruous architecture of the Indian and Portuguese war cemeteries near
Neuve Chapelle. Such jarring architectural intrusions are not of course limited to
the former war-fronts. Witness for instance Melbourne's Shrine of Remembrance;
inspired by the Mausoleum of Halicarnassus and positioned in accordance with
Stonehenge-like astronomical calculations (Moriarty 2009: 146, Sumartojo, this
volume).

These papers also make the point that these were and are *political* spaces. As
Jay Winter pointed out in his seminal work on the subject, 'commemoration was a
political act; it could not be neutral' (Winter 1995: 82). This consideration obliges
us to face some important questions which were posed by the same historian about
memorials; although they are equally relevant to commemorative spaces in their
widest sense: 'Whom or what do they commemorate? Precisely what about the
Great War do they ask us to remember?' (Winter 1995: 78). The centenary of the
war brings these issues into sharper focus, while also reminding us that now might
be the time to step back and analyse the fluid chemistry of almost 100 years of
commemoration.

Creating a language of remembrance

On a cloudless Sunday in June 2006, I found myself on the southernmost fringe of the Alps at Monte Pasubio. During the First World War, this mountain had been an Italian defensive bastion, protecting the vulnerable Venetian plain from Austro-Hungarian attack. Its tactical importance is made all too clear to the visitor by the spectacularly vertiginous view from its flanks – with the industrial town of Schio clearly visible at the head of the plain below, even though a summer haze. As I took in this panorama, I was standing between an ossuary containing the bones of over 5,000 men who perished in the fighting and a museum, newly redeveloped, which recorded the battles fought in the region by the Italian First Army. Thus I was not only standing on a natural commemorative space, but also between examples of the two principal man-made types of commemorative space – a memorial and a museum.

Of the two artefactual spaces, one might naturally assume that the memorial or monument would be the first to appear in the wake of the war; but this was not always the case. For example, the Imperial War Museum existed (at least in committee form) from March 1917, and began its collecting immediately (Cornish 2004). In the same year, the Australian War Records Section began collecting material which would eventually form the basis of the Australian War Memorial – at once a museum and a memorial. So from the outset, memorials and museums were being developed concurrently. This process was an international one, although seldom collaborative and never uniform across nations. Its first flourishing continued for more than 20 years; indeed the Australian War Memorial was not to be installed in its dedicated home until 1941 (Londey and Steel 2008). In a subsequent iteration, from the 1960s onwards, new 'sites of memory' – predominantly in museum form – created 'prosthetic landscapes' (Osborne, this volume).

The original impulse to give physical form to commemoration stemmed from the novel character of the First World War. With whole societies engaged in the war effort, the need was perceived to recognise the contributions of all. Furthermore, death and loss, or the risk of it, was on a scale that touched everyone. As Stéphane Audoin-Rouzeau and Annette Becker (2002: 204–212) argue, in the case of France at least, this was a literal fact. That these morbid circumstances required physical expression is proved by evanescent wartime artefacts like 'street shrines' and *Nagelfiguren*. The groundswell of public interest in commemoration was noted and refined by those in positions of influence. In Britain the editor of the *Connoisseur* magazine, C Reginald Grundy, urged action on the grounds that

> we may fondly imagine that the war in which we are now engaged is so stupendous in its extent, and decisive in its influence on the future destiny of the world, that no important episode connected with it will be forgotten . . . Yet unless we do differently to what we have done in the past this will not be so . . . the names of millions of rank and file, who have served by land or sea, will be forgotten by everyone except their immediate descendants.
>
> (Grundy 1917)

Grundy favoured local war museums but others, closer to the centre of power like Sir Alfred Mond, pushed for a national museum. It was the latter that gained government support in 1917. At almost the same time another leading figure in the history of commemoration, Fabian Ware, secured the creation of the Imperial War Graves Commission, which would subsequently set the tone for the memorialization of Britain's war dead. The Commission appointed three principal architects, Herbert Baker, Reginald Blomfield and Edwin Lutyens. Thus, even before the war's end an official state language of remembrance was instituted.

These influential figures were what might be called political with a small 'P'. The Imperial War Museum was intended not as a celebration of victory, but a record of the war efforts of the people of the Empire. The language of mourning and commemoration which was created by the Imperial War Graves Commission was aimed at reassuring the living that the war dead had 'fallen' in a just cause and could now 'rest' in something approximating a British garden or country churchyard (the design of these cemeteries is now taken for granted, but they are of course further examples of alien intrusions into the existing landscape). In the same vein, Lutyens's Cenotaph in London's Whitehall provided a restrained and surprisingly religion-free focus for official ceremonies of remembrance. Nevertheless, these commemorative spaces *were* political. Just as any museum, according to Jennifer Barrett (Wallis and Taylor, this volume) 'will always carry with it the legacy of its origins, for better or worse', so it is with any memorial or war cemetery. Even local war commemorations, which, across Europe, were the preserve of local councils, service associations, or private individuals were not immune from presenting political messages. For example, Ernst Barlach's extraordinary *Magdeburger Ehrenmal*, or Ernst Friedrich's anti-war museum in Germany; and, in France, local memorials that spoke only of grief and loss, or that were even openly pacificistic (Kidd 2004: 150; Audoin-Rouzeau and Becker 2002: 189). Generally, however, national forms of remembrance set the tone for all. In Britain this encouraged (at least with regard to local memorials) artistic and political conservatism; notwithstanding extravagances like the Loughborough Carillon or the challenging modernity of Eric Kennington's memorial to 24th Division (Black 2004: 137–140). British memorials could even be, in stark contrast to those of continental Europe, utilitarian – meeting-halls or clocks (Audoin-Rouzeau and Becker 2002: 187). Elsewhere, however, the tone could be very political indeed.

Mobilising the dead

This politicization was particularly evident in Italy, where the ossuary of Monte Pasubio (built 1920–26) was the first of a number of sites of remembrance created along the wartime frontlines. In 1921 and 1924, respectively war museums were set up at Rovereto and Gorizia. The locations are significant: both cities stood on conquered territory, so these were more than museums, they were nationalist symbols planted in 'redeemed' parts of *Italia Irredenta*. This was made explicit in the original name of the Gorizia institution: *Museo Della Redenzione* (Museum of the

Redemption). By the time these expressions of Italian nationalism had been completed, the country was in the thrall of Fascism. Under Mussolini the politicization of commemoration was taken further, as the war dead were re-enlisted in support of the Fascist *Weltanschauung*. On the same Alpine front as the existing ossuaries, a huge new one in 'rationalist' style was created at Monte Grappa, followed by a large cemetery at Asiago. The final dramatic flourish was the massive cemetery at Redipuglia on the Isonzo (Soča) front, which was inaugurated in 1938, at the height of the Fascist regime's power. This hillside sheathed in marble holds the remains of 100,000 men – re-mustered from other places of burial to make clear to all the level of sacrifice required if Italy was to attain its destiny (Saunders and Cornish 2014: 6).

This type of politicization of memory continues to echo in the present-day, as commemoration of the war dead remains in use as a means of fostering national identity (Osborne, this volume). It certainly forms part of the function of the great commemorative edifices of Canada and Australia. Although these democratic, multicultural societies are of course utterly different from Fascist Italy, they still deploy the memory of the dead of the First World War as a medium for celebrating the origins of their nationhood, and as a means of binding new generations into the national community (Sumartojo and Foster, this volume). Anzac Day, the anniversary of the first landing at Gallipoli, gained currency in Australia as a commemorative event even during the war, and became the nation's national day. Commemorative spaces like the Shrine of Remembrance and the AWM provide the backdrop for major Anzac Day ceremonies. But even this apparently straightforward patriotic ritual contains a historic political rift within it however, for, as David Stevenson points out, it celebrated only those who volunteered to serve (there was no conscription in Australia) 'and by extension the Protestant and Anglo-Saxon over the Catholic and Irish elements of the population' (Stevenson 2005: 551).

The contested ground on which Italy's war cemeteries and museums were established remained Italian, and there they still stand, but this was not necessarily the case elsewhere. The major German monument to the First World War stood on the site of the Battle of Tannenberg – the 1914 victory over a Russian invasion of East Prussia. In fact, Tannenberg was just one of many towns and villages dotting the site of this huge battle of manoeuvre, but its name was chosen as it suggested revenge for a defeat suffered by the Teutonic Knights in the battle of that name fought in 1410. The 1914 victory had been hugely celebrated in Germany. This was officially encouraged, as the battle represented a war of self-defence that all elements of society could support and drew attention away from the catastrophic failure to win a quick victory in the West during the same autumn. Completed in 1924 this monstrous fortress-like edifice, its walls studded with eight towers, was the resting place of 20 unidentified German dead from the 1914 battlefield. Despite the name given to both the battle and the memorial, it was built near the town of Hohenstein, in the centre of the former battlefield.

Like the Italian Fascists, the Nazis were eager to mobilize this sort of martial commemoration in support of their own extreme nationalism. In August 1933, as the new regime sought to secure its hold on Germany, the party organized a huge rally at the Tannenberg Memorial to mark the anniversary of the battle. In the following year an even more elaborate ceremony accompanied the interment within its precincts of Paul von Hindenburg, the victor of Tannenberg and lately President of Germany. An imposing tomb was created for him within the memorial. In 1935, Hitler designated the Tannenberg Memorial as Germany's national war memorial – a thing that successive post-war governments had failed to create (Rossol 2015). The dictator's subsequent failed attempt to establish German hegemony in Europe meant that Hindenburg's mortal remains lay there only until 1945, when they were removed to keep them from the avenging hands of the Red Army. The retreating Germans then attempted to demolish the memorial, although its complete obliteration only occurred after the area became part of post-war Poland. Ironically, a modern Polish monument to the 1410 battle stands a few kilometres away, just outside the village of Stębark, formerly known as Tannenberg.

This was perhaps the most dramatic fate to meet a politically contested commemorative space, but there are other examples of destruction. The Butte de Warlencourt, a pimple of chalk at the limit of the British Army's 1916 advance in the Battle of the Somme, began its existence as a monument of much earlier people, being an ancient burial mound. From 1917 onwards it was turned into a memorial site by whichever army had control of it – first the British, then the Germans, then the British again. Each replaced the other's monuments with their own, culminating in a re-establishment of German 'control' during the Second World War. This multi-layered site is now firmly defined as a British memorial once again, under the ownership of Britain's Western Front Association (Saunders 2004: 10–11). Memorials have also been eradicated for reasons of internal politics, rather than war. Pacifist memorials erected by Italian Socialists were suppressed by the authorities, or attacked and destroyed by Mussolini's black-shirted *Squadristi* (Janz 2015). Barlach's memorial was removed from Magdeburg cathedral under the Nazis, as was his (less overtly political) hovering angel from the church in his hometown of Güstrow.

To step briefly outside the confines of 1914–1918 and its commemoration, there are other sites of memory that remain actively contested until this day. In Spain, Franco's Valle de los Caídos – a triumphalist nationalist cemetery and memorial masquerading as a symbol of national reconciliation – poses a political problem that an expert commission and governments of differing complexions have failed to resolve. Descendants of Republicans buried there have campaigned to have their bodies removed, while the presence on the site of the remains of Falangist leader José Antonio Primo de Rivera and of Franco himself divide the political Left and Right. In Italy, a less well known, but similarly aligned, political battle is waged over a 180m deep sinkhole in Friuli known as the Bus de la Lum. Those who honour the memory of the Fascism claim that hundreds of captured Fascists were murdered and thrown into this cavern by Communist partisans. They have erected a monument and hold ceremonies to memorialise this alleged event, despite the fact that speleological investigations in the 1950s found the bones of only 28 individuals (De Nardi 2016: 87–92).

The mutability (temporal and political) of engagements with memorials occasionally takes on a personal tone. In the nineteen-sixties Sir Douglas Haig's portrait in Brasenose College was subtitled by students 'Murderer of One Million Men' and the college war memorial was removed (Bond 2002: 54). In 1998 the *Daily Express* newspaper called for the removal and melting down of the same Field Marshal's equestrian statue in Whitehall. Conversely, new memorials can be raised to those who, with the passage of time, come to be considered worthy of them. For example, the 'Shot at Dawn' memorial erected at Britain's National Memorial Arboretum in 2001, which commemorates the 306 British servicemen executed for desertion or cowardice during the First World War. The sculptor of the statue of a blindfolded soldier awaiting execution based the figure on a real person: 17-year old Private Herbert Burden (Black 2004: 144–147, Wilson, this volume). A further example of public investment in the memory of a specific dead serviceman is the case of French writer Alain-Fournier (author of *Le Grand Meaulnes* – beloved of generations of French readers). He was killed in action in Lorraine in September 1914. In 1964 a memorial was erected to his memory, but his grave remained undiscovered; leading to rumours that he had been captured and executed. In 1991, an archaeological investigation found and exhumed his remains and a second memorial was built on the burial site (Balbi 2011: 222–223).[1]

Whose memory? Whose memories?

Commemorative spaces are frequently highly political without being overtly conflicted or subject to appropriation by political causes. For example, in Alsace and parts of Lorraine the usual war memorial inscription 'Mort pour la France' is replaced by 'A Nos Morts', because the dead commemorated by them are most likely to have perished while fighting in the German Army. By the same token, the image of the uniformed *Poilu*, so common on war memorials elsewhere in France, is eschewed in favour of naked figures or representations of grieving women. The same circumstances – a post-war border change – produced an analogous situation with regard to memorials in Italy's Trentino-Alto-Adige province and in parts of Venezia Giulia, which commemorate men who fought in the Austro-Hungarian armed forces. Jay Winter's two questions are brought inexorably to mind.

In many cases, the politics of commemoration is manifested as much by the absence of its physical expression as by its presence. The geography of commemorative spaces is lopsided, because different communities or political consensuses engaged with war-memories in different ways. Most obviously those on the losing side, effectively Germany and post-1918 Austria, were commemorating losses suffered in a defeat. The German national focus on the victory of Tannenberg was just one outcome. Proposals for a national war memorial became mired in regional and factional issues; meaning that it took until 1931 to even agree a site (Rossol 2015; Goebel 2015). As a defeated nation (and one which suffered two major economic crises) Germany did not enjoy the same opportunities as the victors for interring and memorializing its war dead. The German war cemeteries in France were cramped, with many multiple graves, and mourning families were not able to visit them freely until 1925 (Watson 2015: 565). In Austria, commemoration was focused on the losses suffered by the German-speaking people of the once vast Hapsburg Empire. For the peoples of the ethnically based states that rose from the wreck of Austria-Hungary, the war was no longer perceived as 'theirs' – or at least only as the mechanism by which their independent existence had come about. This was despite the considerable numbers of them who had fought in the ranks of the Empire's armies, many (particularly, for example, Slovenes, Bosnians and Poles) with wholehearted effort. Now, if commemorated at all, the dead were interpreted as having perished in the struggle for independence (Watson 2015: 565). For the peoples of these regions, the recent past was – in a literal sense – 'a foreign country'. In the newly created Soviet Union, memory of the First World War was largely expunged, not just because of the adoption of a new political identity, but because the struggle itself had been overshadowed by a far greater disaster in the form of civil war. In such a political climate it was quite possible for major sites of commemoration to simply disappear under vegetation over a period of years (Zalewska 2016: 157–162).

Such 'absence' was not unknown in the British Empire, where even the lofty ideals of the Imperial War Graves Commission could be watered-down for reasons of expedience. Black Africans were largely excluded from receiving the supposedly universal standard of interment and memorialization accorded to the British Empire's war dead (Barrett 2014: 80–90) The decision might have been prompted by financial considerations, but the arguments propounded to justify it take us into the politics of race and of racism. Only in this century has Britain begun to grapple with the issue of commemorating the role played in the Great War by its then subjects in Africa, Asia and the Caribbean (Wilson, this volume). For a post-First World War example of this sort of asymmetric memorialization, Spain once again springs to mind; with many relatives of Republican victims of the nationalist *Limpieza* denied not only memorials, but even precise locations on which to focus their mourning (González-Ruibal 2014: 169–181). Absence is also apparent in the commemoration of dissenting voices. Conscientious objectors, who represented a significant socio-political response to Britain's 1916 decision to introduce compulsory military service, were not commemorated until 1994, when a stone dedicated to them was unveiled in London's Tavistock Square (Wilson, this volume). Its location was the outcome of London-based fundraising and the willingness of Camden Council to accommodate it. But a more suitable spot might have been found in the valleys of West Yorkshire, where opposition to conscription was more strongly concentrated than anywhere else (Pearce 2001). A similar absence of physical commemoration might said to be evident at the site of the most concerted political opposition to the war itself: 'Red' Clydeside. It is evident nevertheless that the imprint of memory left on the area by the Great War has influenced the outward manifestations of modern-day anti-war activism in the same area (Griffin, this volume).

A new wave of memorialization of the First World War commenced in the 1960s and, in Britain at least, now appears to be reaching a crescendo in the opening decades of the twenty-first century (Wilson, this volume). Among its earlier manifestations were new museums, the missions and content of which reflected changing attitudes to the war itself. At the same time, existing museums (such as those in Italy) were adopting more nuanced and inclusive takes on the war. In France, the re-positioning of the country in the post-1945 world – especially in the context of its partnership with its former deadly rival, Germany – predicated a museological engagement with the First World War which was avowedly not nationalistic and which fostered reconciliation. Its first expression was in the 1967 *Mémorial de Verdun*. In more recent times, the *Historial* at Péronne (1992) and the *Musée de la Grande Guerre* at Meaux (2011) have been born out of a happy coincidence of existing private collections, local political will and the availability of funding. Both adopt a modern, academically based approach to their subject.[2] Belgium can boast an equivalent in the In Flanders Fields Museum at Ieper, which has gone through two successful iterations (1998 and 2012).

Perhaps even more significant (at least in the context of the absence of commemoration) is the creation of a museum at Kobarid in Slovenia, dedicated to the battle to which that town (formerly Karfreit or, to Italians, Caporetto) has lent its name. Opened by local enthusiasts in 1990, this museum gained official status following Slovenian independence. It is emblematic of an awakening of interest in the First World War in a region where it had long lain dormant – despite the fact that the wartime front line lay largely in its territory and that Slovenians had fought hard to defend their land as soldiers of the Hapsburg *Kaiserliche und Königliche Armee* and *Landwehr*. A similar recognition – through a site of memory – of a war formerly dismissed as someone else's can be traced in Senegal, where in 2004 it was decided to institute a memorial day for African soldiers who had died in both World Wars. This led to the return to the centre of Dakar of a French First World War memorial. Featuring the figures of two victorious soldiers, one French, one Senegalese, this statue had previously been removed from a prominent position in the capital for being emblematic of French colonialism. In 2007 a further example of renewed African engagement saw an imposing memorial tower, dedicated to the war dead of Malawi, erected in the capital Lilongwe (Chetty and Ginio 2015). Memorials built by the British colonial administrations in the East African cities of Nairobi, Mombasa and Dar es Salaam – all featuring statues of indigenous servicemen – have, by contrast, remained in place since their inauguration (Waller 2009: 88–89).

Politics of a social variety are in evidence when it comes to deciding who is eligible to visit and engage with the commemorative spaces of the First World War. This has been the case since the very earliest stage of their development – or even before. Tour operators were at pains to state that battlefield visits would not be conducted until after the war's end (Pennell, this volume). When such tours did begin, they were publicized as 'pilgrimages' by British companies such as Thomas Cook. This echoes a concern expressed by Lieutenant Colonel Rowland Feilding in a 1918 letter to his wife. After showing a souvenir-hungry American around the old Loos battlefield, he reflected 'It is horrifying to see this sacred ground desecrated in this way, and still more to think of what will happen when the cheap tripper is let loose. With his spit he will saturate the ground that has been soaked with the blood of our soldiers' (Feilding 1929: 367). On a visit framed as a 'pilgrimage', however, it might be anticipated that participants would repress urges to take souvenirs or expectorate. Incidentally, contemporary posters indicate that a less reverent and more pragmatic attitude was the norm in France and Belgium, where visits to the former front line were marketed as jaunts which could be undertaken by individuals by train or motor-car (see IWM posters Art.IWM PST 12773; Art.IWM PST 3951). The money spent by tourists was welcome in areas devastated by war – as evidenced by the mass of material produced locally for sale to them (Saunders 2003: 49–51, 143–152).

Depredations by souveniring visitors could actually prompt efforts to create official, and therefore controllable, commemorative spaces. The memorial of the *Tranchée des bayonettes* at Verdun was partly an effort to protect this site of patriotic legend from despoliation by relic hunters (Winter 1995: 99). In the early years of the Imperial War Museum considerable efforts – including the deployment of barbed wire – had to be made to protect exhibits from damage by acquisitive visitors; much to the vexation of museum secretary Charles ffoulkes, who regarded his collections as 'sacred relics' (Cornish 2004: 42–43, 46). Such concerns continue into modern times: witness the 2015 case of two British schoolboys who pilfered relics on a visit to Auschwitz (Guardian 2015). Unsurprisingly, given this context, it would appear that some teachers select pupils for visits to Western Front commemorative sites on the basis of their likely behaviour (Pennell, this volume). So selectivity, in various forms, has long been applied in deciding who might be considered fit to engage with commemorative spaces. This can of course include self-selection – for, as one prominent figure in the town of Ieper is reported as saying, 'the English people who ought to come here don't' (Hurst 2013: 41).

This afterword has been something of a canter around a selection of the world's monuments to the First World War. That selection is not by any means comprehensive of course, but I believe that it yields sufficient evidence to make clear that commemorative spaces are inherently political (in all senses of the word) in nature. As manifestations of the material culture of conflict they live a 'social life', which is engendered by the relationship that people have with them. Much of this engagement is political and, as a result, these spaces are frequently fiercely contested. In them we see the social construction of landscape at work and, as with all manifestations of material culture, the passage of time leads to the application of new and differing layers of engagement (Osborne, this volume, Wilson, this volume). These influences leave us with what Nicholas Saunders has described as 'a complex palimpsest of overlapping, multi-vocal landscapes' (Saunders 2004: 7). This brief survey, stimulated by the varied papers given at the 2014 RGS-IBG Annual International Conference, has looked back over a century of the social lives of commemorative spaces. By the time it is published those lives will have moved on again and new layers of meaning will have been deposited. The revision, renegotiation and reconstruction of our engagement with the commemorative spaces of the First World War will continue as long as they, or we, exist.

Notes

1 At the same time the dig in question established the remains of the First World War in France as a legitimate subject for archaeological investigation and as a cultural patrimony worthy of legal protection.
2 For a detailed description of the former's origins and philosophy see Winter (2006: 222–237).

Bibliography

Audoin-Rouzeau, S. and Becker, A. (2002) *1914–1918: Understanding the Great War*, London: Profile.

Balbi, M. (2011) 'L'Archeologia dei Nonni', in F. Nicolis, G. Ciurletti and A. De Guio (eds.), *Archeologia della Grande Guerra*, Trento: Provincia autonoma di Trento.

Barrett, M. (2014) '"White Graves" and Natives', in P. Cornish and N.J. Saunders (eds.), *Bodies in Conflict*, London: Routledge.

Black, J. (2004) 'Thanks for the memory: War memorials, spectatorship and the trajectories of commemoration, 1919–2001', in N.J. Saunders (ed.), *Matters of Conflict*, London: Routledge.

Bond, B. (2002) *The Unquiet Western Front*, Cambridge: Cambridge University Press.

Chetty, S. and Ginio, R. (2015) 'Commemoration, cult of the fallen (Africa)', in *1914–1918 Online: International Encyclopedia of the First World War*. Available from: http://encyclopedia.1914-1918 online.net/article/commemoration_cult_of_the_fallen_africa.

Cornish, P. (2004) '"Sacred relics": Objects in the Imperial War Museum 1917–1939', in N.J. Saunders (ed.), *Matters of Conflict*, London: Routledge.

De Nardi, S. (2016) *The Poetics of Conflict Experience: Materiality and Embodiment in Second World War Italy*, London: Routledge.

Feilding, R. (1929) *War Letters to a Wife*, London: The Medici Society.

Goebel, S. (2015) 'War Memorials (Germany)', in *1914–1918 Online: International Encyclopedia of the First World War*. Available from: http://encyclopedia.1914-1918-online.net/article/war_memorials_germany.

González-Ruibal, A. (2014) 'Absent bodies: The fate of the vanquished in the Spanish Civil War', in P. Cornish and N.J. Saunders (eds.), *Bodies in Conflict*, London: Routledge.

Grundy, C.R. (1917) *Local War Museums: A Suggestion*, London: Bemrose & Sons.

Hurst, S. (2013) *Ypres, The Great War and the Re-Gilding of Memory*, Ypres: In Flanders Fields Museum.

Janz, O. (2015) 'Mourning and cult of the fallen (Italy)', in *1914–1918 Online: International Encyclopedia of the First World War*. Available from: http://encyclopedia.1914-1918-online.net/article/mourning_and_cult_of_the_fallen_italy.

Kidd, W. (2004) 'The lion the angel and the war memorial: Some French sites revisited', in N.J. Saunders (ed.), *Matters of Conflict*, London: Routledge.

Londey, P. and Steel, N. (2008) 'Der erste Weltkrieg als nationaler erinnerungsort: das Imperial War Museum in London und das Australian War Memorial in Canberra' (The First World War as a site of national memory: The Imperial War Museum in London and the Australian War Memorial in Canberra), in W. Hochbruck, B. Korte and S. Paletschek (eds.), *Der Erste Weltkrieg in der populären Erinnerungskultur*, Essen: Klartext-Verlag.

Moriarty, C. (2009) 'The returned soldier's bug', in N.J. Saunders and P. Cornish (eds.), *Contested Objects*, London: Routledge.

Pearce, C. (2001) *Comrades in Conscience*, London: Francis Boutle.

Rossol, N. (2015) 'Commemoration, cult of the fallen (Germany)', in *1914–1918 Online: International Encyclopedia of the First World War*. Available from: http://encyclopedia.1914-1918-online.net/article/commemoration_cult_of_the_fallen_germany.

Saunders, N. (2003) *Trench Art*, Oxford: Berg.

Saunders, N. (2004) 'Material culture and conflict', in N.J. Saunders (ed.), *Matters of Conflict*, London: Routledge.

Saunders, N. and Cornish, P. (2014) 'Introduction', in P. Cornish and N.J. Saunders (eds.), *Bodies in Conflict*, London: Routledge.

Stevenson, D. (2005) *1914–1918*, London: Penguin.

The Guardian (2015) [Online]. Available from: www.theguardian.com/world/2015/jun/23/uk-teenagers-held-thefts-artefacts-auschwitz-museum (accessed 23 June 2015).

Waller, R. (2009) 'Subversive material: African embodiments of modern war', in N.J. Saunders and P. Cornish (eds.), *Contested Objects*, London: Routledge.

Watson, A. (2015) *Ring of Steel*, London: Penguin.

Winter, J. (1995) *Sites of Memory, Sites of Mourning*, Cambridge: Cambridge University Press.

Winter, J. (2006) *Remembering War*, New Haven: Yale University Press.

Zalewska, A. (2016) 'The "gas-scape" on the Eastern Front, Poland (1914–2014): Exploring the material and digital landscapes and remembering those "twice-killed"', in B. Stichelbaut and D. Cowley (eds.), *Conflict Landscapes and Archaeology from Above*, Farnham: Ashgate.

Index

Page numbers in italics indicate figures and tables.

Milton Keynes UK
Ingram Content Group UK Ltd.
UKHW040107071024
449327UK00019B/882